PROCESS MEASUREMENT AND CONTROL

Introduction to Sensors, Communication, Adjustment, and Control

Roy E. Fraser
Humber College of Applied Arts and Technology

Upper Saddle River, New Jersey
Columbus, Ohio

Library of Congress Cataloging in Publication Data

Fraser, Roy E.
Process measurement and control : introduction to sensors, communication, adjustment and control / Roy E. Fraser.
p. cm.
Includes bibliographical references and index.
ISBN 0-13-022211-9
1. Process control--Instruments. 2. Measuring instruments. 3. Detectors. 4. Automatic control. I. Title.

TS156.8. F72 2001
629.8--dc21

00-027088

Vice President and Publisher: Dave Garza
Editor in Chief: Stephen Helba
Assistant Vice President and Publisher: Charles E. Stewart, Jr.
Production Editor: Alexandrina Benedicto Wolf
Production Coordination: Tim Flem/PublishWare
Design Coordinator: Robin G. Chukes
Cover Designer: Mark Shumaker
Cover Image: Stock Market
Production Manager: Matthew Ottenweller
Marketing Manager: Barbara Rose

This book was set in Times Roman by PublishWare. It was printed and bound by R. R. Donnelley & Sons Company. The cover was printed by Phoenix Color Corp.

10 9 8 7 6 5 4 3 2
ISBN 0-13-022211-9

Olgai, manai mīļai dzīvesbiedrei

And to our wonderful family

Martin, Jane, and children
Laura, Valerie, and Melinda

Natalie, David, and children
Daniel, Erin, and Tessa

PREFACE

This book is a text for students learning automatic control systems and a reference for those pursuing careers in industry. When readers select a measuring sensor or a control valve, this book will refresh them about the important details that must be considered. When they interconnect the sensor and control valve into a communicating network, this book will remind them of the important methods for minimizing interference from electrical and magnetic fields. When they tune a controller, it will guide them toward establishing a desirable control loop response to a disturbance.

The concepts discussed in this book focus on a classic process control loop with its four major blocks: the measuring sensor block, the controller block, the process adjustment block, and the process block. Design procedures include the selection of measuring sensors and control valves. The sensors concentrate on the measurement of temperature, pressure, flow rate, and level, and analysis of humidity and electrolytic conductivity. Also covered is the design of network wiring to minimize interference.

Operational procedures show the presentation of process information for safely monitoring the trajectory of a changing process (e.g., trend display and alarm status), and they describe control loop responses to disturbances. The financial investment in this kind of equipment is also reviewed.

Maintenance procedures emphasize calibration of measuring sensors. Maintenance of a network to minimize interference from inadvertent grounds on the network is also described. Additionally, the tuning of controllers to achieve a desirable response to disturbances is emphasized.

Each chapter begins with objectives that present the concepts covered. At the end of each chapter are sets of questions and problems, suggested practical lab assignments, and references that provide additional resources on the concepts covered in this book. Selected problems have answers in Appendix C.

Chapter 1 introduces continuous automatic control and differentiates it conceptually from discrete control. The control loop is emphasized as the basic concept for continuous control of a process variable. The four blocks of every control loop are described, and their input and output signals are shown on a block diagram. The Instrument Society of America symbols are presented for describing control systems on a Process and Instrument Diagram (P&ID).

Chapter 2 begins with instrument calibration concepts and documents. Instruments for sensing temperature, pressure, flow, level, humidity, and electrolytic conductivity are described, and many of the device limitations for these measurements are mentioned. These limitations are shown as important elements involved in the selection of a device for a successful application.

The presentation of process control information for human operators and for machines is introduced in Chapter 3. The overview display, process group display, and the point display are described. Additional displays include trend, alarm summary, and hourly averages.

Chapter 4 presents techniques to achieve reliable, secure data communications around the control loop. Comparison of pneumatic and electric communications are made to show the speed of the electrical technique and its potential hazards. Problems due to interference from magnetic and electric fields and from improper electrical grounding are described, along with proper techniques to minimize these problems. The concepts of networks are introduced.

Chapter 5 emphasizes the application of control valves for process adjustment. The sizing of control valves for liquid and gas flow adjustment follows ISA standard S75.01.

Chapter 6 introduces the proportional, integral, and derivative (PID) controller and explains its functions and the ultimate cycle method for its tuning. Ratio control and cascade control of a process are also explained.

The appendixes describe ISA symbols, display blank calibration documents, display photos of control valves, list answers to selected problems, list ASCII symbols and codes, describe calculations for liquid pressure drops in piping, and list a table of thermocouple millivolts versus temperature and a table of RTD resistance versus temperature.

Students and instructors should review the Study Procedures on page vii to assess whether they have the prior training required to rapidly absorb the information in this book. The estimated time required for students to learn the material, study the theory, answer questions and problems, and perform the assignments is also given. The instructor's manual available for use with this text provides an outline of study and further information on how to successfully teach classes in process measurement and control.

Acknowledgments

I wish to thank Humber College for supporting me in the preparation of this book. Colleagues from the Electrical/Control Systems program who encouraged me include Jeff Dixon, Mike Birmingham, and Tarsem Sharma. The ISA–Toronto section was very helpful, especially Len Klochek, who made many suggestions for improvements. I would also like to thank the following reviewers for their valuable feedback: J. Tim Coppinger, Texas A&M University–Corpus Christi; Mohan Kim, San Jose State University; Xiaoli Ma; David Pacey, Kansas State University; William Reeves, Ohio University; Wahija Shireen, University of Houston; and Gang Tao, University of Virginia. My thanks to Alex Wolf of Prentice Hall and Tim Flem and Jim Reidel for significantly improving this book. I wish to thank Charles E. Stewart of Prentice Hall for his confident support.

Roy Fraser
Humber College of Applied Arts and Technology

STUDY PROCEDURES

This book is designed for students planning careers as measurement and control technicians, technologists, and engineers. At a minimum, such students should have had mathematics background, including algebra, trigonometry, plane geometry, and complex variables. They should also have studied electricity and electronics to the point where they can analyze circuits for current, voltage drops, and impedances, and can use ammeters, voltmeters, and oscilloscopes to troubleshoot a circuit. A basic expertise in using computer spreadsheets, word processors, and CAD programs is also required.

This book is intended for a structured lecture program, but it may also be used in an individualized-learning program with equipment for lab assignments. The lab assignment demonstrates to the student the concepts described in the lesson and is closely related to the industrial application of those concepts.

To completely learn the material in this book, the student needs to invest approximately two lecture hours, two lab assignment hours, and two to four home hours—to study before the lecture, to do the problems after the lecture, and to prepare for the lab assignment and report—each week during a 16-week semester. A recommended lab assignment report format is shown below. About 30 hours are required in the classroom, 30 hours in the lab, plus another 60–70 hours study time.

Project or Lab Assignment Report Format

Concise, complete reports demonstrate an understanding of the material in the section being studied. The following format will help ensure such reports are produced.

Title Page

- Course name and number
- Assignment number and title
- Instructor's name
- Date assignment was performed
- Student's name
- Partner's name

Subsequent Pages

Objective	Include a copy of the stated objective.
Procedure	Include a copy of the requested procedure to follow.
Diagrams	Show how you actually connected the equipment.
Data	List the data that you collect.
Results	Describe your actual results and compare them to expected results. Describe any unexpected results.
Conclusions	Answer any questions. Try to give reasons for unexpected results.

An effective report will help the student better understand the subject matter, and it will serve as a reference for further work. The report should be started while preparing for or while performing the assignment. The student should have the instructor assess it, initial completed sections, and then mark it within one or, at most, two weeks of the assignment period. Excessive neatness is not essential (simply cross out any incorrect or unnecessary sections). However, the student should keep and organize all the completed reports in a single binder as he or she proceeds through the course.

CONTENTS

1

INTRODUCTION TO PROCESS MEASUREMENT AND CONTROL

OBJECTIVES

When you complete this chapter you will be able to:

- Define process measurement
- Define process control
- Calculate simple return on investment from a process control system
- Sketch a block diagram of a process control loop
- Describe typical industrial processes under process control
- Draw simple Process and Instrumentation Diagrams (P&ID) using ISA symbols
- Describe the measuring sensor block of a control loop
- Describe the controller block of a control loop
- Describe the process adjustment block of a control loop
- Describe the signals circulating around a control loop

1.1 PROCESS MEASUREMENT AND CONTROL DEFINED

Processes include anything from the heating of your house to the marketing of baby food. For our purposes, however, we will be concerned only with industrial processes such as the distillation of crude oil, or the digestion of wood chips to make pulp, or the conversion of pulp to paper, or the fabrication of plastic products such as 1-liter plastic soft drink bottles. These are overall processes and each of them will usually include many subprocesses.

Process measurement is defined as the systematic collection of numeric values of the variables that characterize a process to the extent that the process control criteria of the process user are satisfied. As an example, if you require your house temperature to be maintained between 18°C and 24°C with an accuracy of 0.25°C, then your thermostat must measure the temperature and collect its numeric value for your furnace controller so that it will maintain this accuracy. As another example, if the owner of a distillation plant making gasoline requires a certain octane range from the plant, then all the measuring and control instruments on the plant must be chosen to work together accurately to ensure that the plant achieves those criteria. As a final example, if the manufacturer of baby food wishes to make mashed carrots for the market, then he will ask a market analyst to acquire data on the potential customers of mashed carrots for babies.

The purpose of process measurement is to assist a human or a machine to monitor the status of a process as it remains at some steady state or as it changes from one state to another such as heating up crude oil. In most cases the human or machine will guide or force the process to change safely from an initial state to a more desirable state (e.g., forming a plastic bottle). This is process control.

1.2 INVESTMENT

Of course, the purpose of any process equipment and its measurement and control system is to provide a satisfactory return on its invested capital. The best control system maximizes the return on the whole process plant, not just on the capital invested in the control system. It does this over its lifetime. Therefore, maintenance costs and loss of production costs due to breakdowns should be considered when selecting the measurement and control system. Reliability is a very important aspect in the selection of systems and often worth the added initial costs. Reliability of the system is in itself important. So is the reliability of the system supplier to provide replacement parts and service in the future.

An example of justifying the expenditure of $2,353,000 on a replacement control system for an existing process, using an old, manual method of control, has the following simple, estimated costs:

Annual losses from old control system	
Old system losses due to off-spec product	$850,000
Old system losses due to control system downtime	145,000
	$995,000
Annual savings from new control system	
New system losses due to off-spec product	$250,000
New system losses due to control system downtime	145,000
	$395,000
Reduction of losses due to new system	$600,000
Extra annual operating and maintenance cost of new system	115,000
Annual savings from new system	$485,000

One-time costs of new system	
Lowest price vendor's price	$1,875,000
Spare parts costs	125,000
Installation costs	265,000
Training costs for plant personnel	88,000
	$2,353,000

$$\text{Expected return on investment} = \frac{\$485,000/\text{yr} \times 100\%}{\$2,353,000} = 20.6\%/\text{yr}$$

$$\text{Payback period} = \frac{\$2,353,000}{\$485,000/\text{yr}} = 4.85 \text{ years}$$

This example shows in a simple way the method of comparing engineering projects in order to select the ones that will make the most profit for the company planning to make improvements to its plant. For example, if the company has many possible projects to invest its money in, it will choose to rank them according to their return on investment. The ones with a high return on investment will usually be chosen over the ones with low return on investment.

1.3 CONTINUOUS AND DISCRETE PROCESS CONTROL LOOPS

There are two main forms of process control: continuous and discrete (or on/off). The continuous form of process control implies smooth even measurement of the process variable over a fairly wide range that is finely resolved into many hundreds or thousands of values. For example, the temperature in your home may cover a continuous range of 0°C to 40°C resolved into at least 256 values, or spacings of 0.156°C. The discrete form of process control implies a discrete variable with only a few (at least two, such as on and off) status points covering its range. For example, an electric motor may be stopped (off) or running (on). This book concentrates on continuous control.

Continuous process control emphasizes feedback using a closed loop. Figure 1.1 shows a typical closed loop. An example of a closed loop is cruise control on an automobile. The driver sets a certain desired speed as the set-point signal and places the car on cruise control. If the feedback signal (speed of the car) is less than the set-point signal, and a positive error (difference between set-point speed and feedback speed) exists, then the controller increases its output signal. The output signal opens the throttle (the process adjusting device) of the engine (the process), causing the car to speed up. The speed sensor detects this increase in speed (the measured process variable) and sends a stronger feedback signal to the controller. This action continues until the feedback signal equals the set-point signal, and then the controller output signal maintains its value, keeping the car at the desired speed.

Discrete process control emphasizes Boolean logic using gates and timers. Figure 1.2 shows a generalized logic diagram. An example of discrete process control is

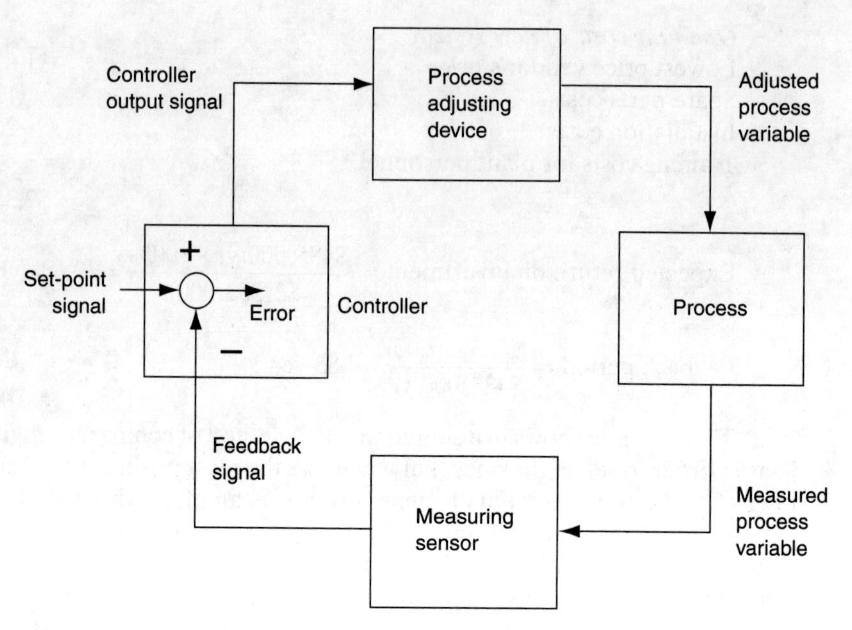

FIGURE 1.1
Typical closed loop for continuous process control

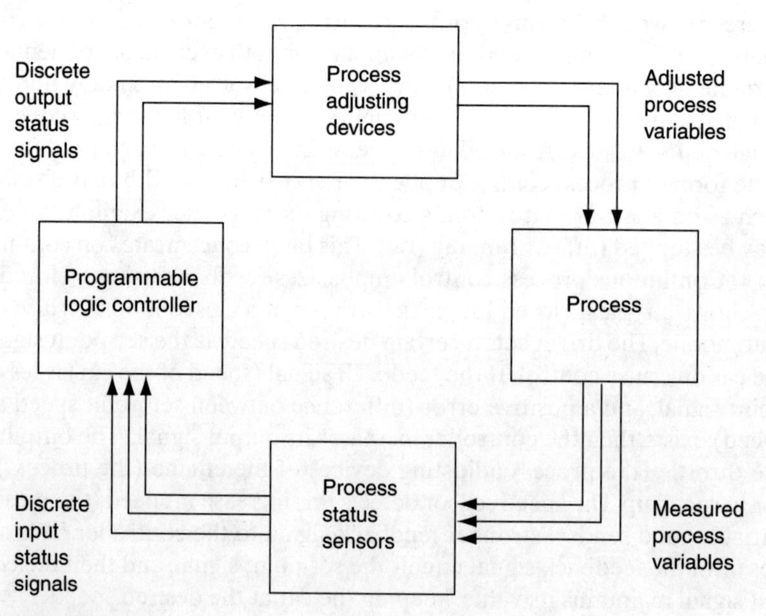

FIGURE 1.2
Generalized logic diagram for discrete process control

the control of a crossing gate at a railroad crossing. Two train sensors are used on the railway tracks. One is placed on the tracks at one side of the crossing and the other on the tracks on the other side of the crossing. If either sensor detects a train, the crossing gate on the highway should close. However, if a short train appears it may be between the sensors and the crossing gate may open at an incorrect time. The control logic should not let this happen; it should only allow the gate to open after the train passes over the second sensor. In this case the process is the flow of traffic over the highway and over the railway. The railway train sensors are the process status sensors, and they produce the discrete input status signals for the programmable logic controller. Based on these signals, the programmable logic controller decides when to change the discrete output status signal. This signal operates the motor (process adjusting device) driving the gate open or closed.

1.4 THE PROCESS BLOCK OF A CONTROL LOOP

Figure 1.1 is a most important figure. The four blocks and the signals that connect them together will be referred to frequently. The most important block is the process block because it dictates how the other blocks are expected to perform. The measuring sensor must be selected to measure the process variable that is associated with the process. The process adjusting device must be selected to adjust the adjusted—or manipulated—process variable that is associated with the process.

The adjusted or manipulated variable adjusts the process so that the measured process variable approaches more closely to the set-point value. For example, in an automobile on cruise control, the throttle position is the adjusted process variable, and it adjusts the fuel flowing to the engine. As the fuel flow is increased, the speed of the auto increases. The speed of the auto is the measured process variable, and if the control system is functioning correctly, the speed should be approaching more closely to the set-point value.

In order to describe control loops, ISA symbols (described in Appendix A) have been used by many large industrial companies for more than 50 years. Each instrument or instrument function is identified with a circle (or "balloon"), some letters, and a number as its symbol. Most of the letters are defined by ISA, but some may be defined by the user. The process symbol is not usually defined by ISA and here the user may be creative. Every control loop has hardware or software that is represented by the four blocks shown in Figure 1.1. For each loop a process and instrumentation diagram (P&ID) is prepared. For example, as shown in Figure 1.3, there are usually many loops shown on one large diagram for a major process. Each loop on the P&ID uses ISA symbols to show the particular devices that perform the functions shown in Figure 1.1. Figure 1.3 shows a typical liquid flow control loop (F-212) supplying liquid from a pipe to a tank. Figure 1.3 also shows a typical liquid level control loop (L-141) adjusting the flow out of the tank to maintain the tank level at the set point of the level controller (LIC-141). By studying Appendix A you should be able to conclude that the letters and numbers associated with each ISA symbol represent the following:

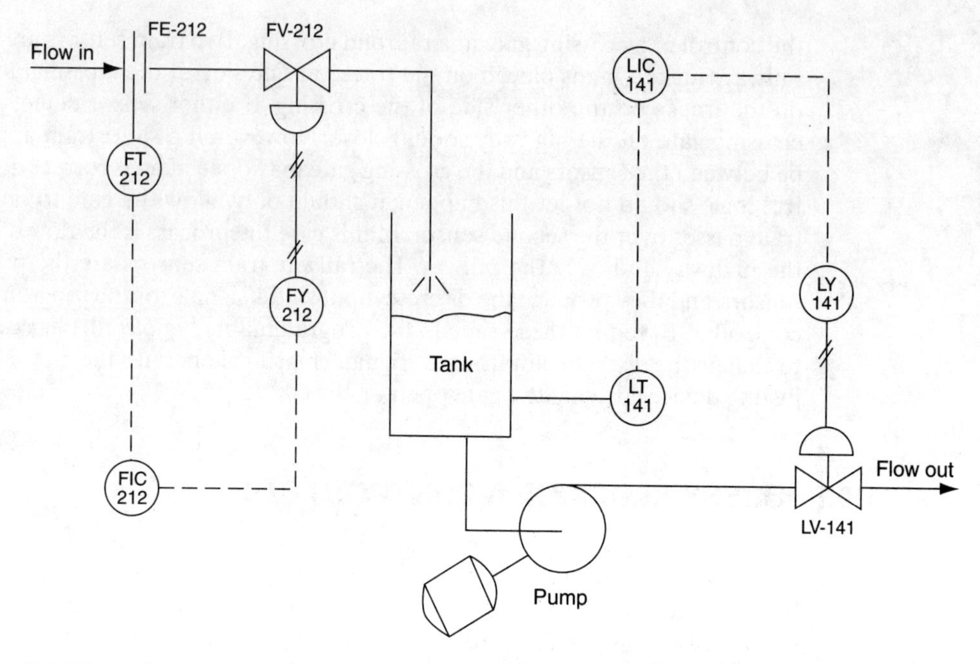

FIGURE 1.3
Typical liquid flow and liquid level control loops

Flow Loop F-212

FE-212	Flow primary element (e.g., orifice plate or venturi)
FT-212	Flow transmitter
FIC-212	Flow indicating controller
FY-212	Flow relay or compute function (computes pneumatic signal to correspond to electric signal value)
FV-212	Flow valve (adjusts flow to correspond to pneumatic signal value)

Level Loop L-141

LT-141	Level transmitter
LIC-141	Level indicating controller
LY-141	Level relay or compute function (computes pneumatic signal to correspond to electric signal value)
LV-141	Level valve (adjusts flow to correspond to pneumatic signal value)

1.5 THE MEASURING SENSOR BLOCK OF A CONTROL LOOP

The measuring sensor block of flow loop F-212 in Figure 1.3 includes both the flow element and the flow transmitter, whereas the measuring sensor block for level loop L-141 needs only the level transmitter. Depending on the process, the block may include several instrument functions. The P&ID shows all the instrument functions so that the detailed design drawings will reflect the complete loop and so that any engineering cost estimates will be complete.

Measuring sensors are described for many of the common process variables in Chapter 2. However, there is an enormous number of sensors for the less common type of process variables that are not described. If you become deeply involved with specifying measuring sensors, then you will want to obtain some of the reference texts listed at the end of Chapter 2.

Notice the dotted line coming from the flow and level measuring sensors in Figure 1.3 to their controllers. This represents the electrical feedback signal (corresponding to the measured process variable) that is transmitted from the measuring sensor to the controller. The most important feature of a measuring sensor is the relationship of the feedback signal to the measured variable. Measuring sensors must maintain this relationship in order to achieve an accurate calibration.

1.6 THE CONTROLLER BLOCK OF A CONTROL LOOP

The controller block of the control loop may be hardware or software. Hardware type electronic controllers receive a feedback signal value from the measuring sensor and compare it with the set-point value (usually dialed in to the hardware by the process operator) to obtain an error value. If the feedback signal has the same value as the set point, the error is zero, and this means that the measured variable equals the set point. The error signal generates an output signal from the controller that is the signal sent to the process adjusting block. Generally there may be three functions with gain settings that make up the controller output signal. These functions are the error itself, its derivative, and its integral. Whether the controller block is hardware or software, it calculates its output signal in the same way. Be careful not to assume that the controller output signal is proportional to the feedback signal. Even though they may have the same range (4–20 ma DC) they are related in a fairly complicated mathematical way.

1.7 THE PROCESS ADJUSTMENT BLOCK OF A CONTROL LOOP

The process adjustment block manipulates or adjusts a process variable in order to change the measured process variable to bring it closer to the set point. For example, in flow loop F-212 of Figure 1.3, the adjusted process variable is the measured process variable, flow. In level loop L-141, however, the adjusted process variable is the flow out of the tank, which is quite different from the measured process variable, the tank liquid level.

Chemical-type process control loops usually adjust fluid flow with control valves. These are commonly the process adjusting device. However, there are many other types of process adjusting devices, including electric motors, electric heaters, pneumatic pistons, pumps, fans, and so on.

1.8 THE SIGNALS CIRCULATING AROUND A CONTROL LOOP

Each block of a control loop has an input signal and an output signal. The input signal of one block is the output signal of another block. The units of a signal may be degrees Celsius or milliamperes or some other physical units. If a range for the feedback signal to the controller block, such as 4–20 ma DC, is settled upon, however, then this may also be considered as 0–100%.

If we look at the flow loop F-212 in Figure 1.3, then we can imagine a typical steady-state gain (an output signal corresponding to an input signal) of each block around the loop. For example:

Instrument	*Input Signal Range*	*Output Signal Range*
FE-212	0–100 GPM	0–100 in. of water
FT-212	0–100 in. of water	4–20 ma DC
FIC-212	4–20 ma DC	4–20 ma DC
FY-212	4–20 ma DC	3–15 psig
FV-212	3–15 psig	0–85 GPM

If we were to open the loop at the feedback signal and insert a sine wave signal to the controller for the feedback signal, then we will see a corresponding sine wave signal appear at the feedback signal from the measuring sensor. If this sine wave has an amplitude in phase with, and greater in value than, the signal injected into the controller, then we obviously have an unstable loop. The way these signals interact with one another establishes the way our closed control loop will perform. Any oscillation should dampen down quickly to a steady state value. For each loop, the possibility of unstable operation must be considered, and the loop must be tuned to esnure stable operation.

PROBLEMS AND LAB ASSIGNMENTS

1.1 A company's process makes a product that is completely sold out each year no matter how much it produces. With the simplest manual control, there are average annual losses totaling $1,260,000 due to off-spec product or downtime due to control system faults. With a sophisticated distributed control system, these losses could be reduced to $225,000. This new control system is quoted by the lowest cost vendor as $2,875,000 plus $260,000 for recommended spare parts. The consultant estimates installation costs to be an additional $470,000. Training of the operator crews and maintenance crews is estimated to be $185,000. The extra annual maintenance costs are quoted as $220,000. What is the annual percent of return on the invested capital for the control system, and what is the payback period?

1.2 Describe the logic that you would use to keep the railroad crossing gate closed in the discrete process control example associated with Figure 1.2 when a short train is between the two train sensors.

1.3 Identify the control loop block associated with each ISA symbol shown in Figure 1.3.

1.4 Draw a P&ID diagram using ISA symbols for a temperature control loop that controls the temperature of the liquid in the tank of Figure 1.3 using a steam-heating coil located in the tank liquid.

1.5 **Continuous (or Analog) Control Loop Identification and Operation**

Objective: To be introduced to ISA symbols, to identify the features of a continuous process control loop, and to operate the loop in the automatic (closed loop) mode and manual (open loop) mode.

Equipment: A demonstration process control loop using a commercial analog controller. The preferred system is a liquid level loop with a plexiglass tank tower about 6 or 8 inches in diameter and 50 inches high.

Alternatively a computer simulation of a tank under automatic level control, such as TANKSIM, may be used.*

Identification Procedure: Use an electronic or pneumatic liquid level control loop and identify all the features shown in Figure 1.1. This loop should have its control valve on the inlet to the tank and a manual valve adjusting the flow of liquid (water) out of the tank to a drain. Alternatively, the manual and automatic control valves may replace one another. Draw the loop using ISA instrumentation symbols (see Appendix A). Prepare a list including each device with the ISA tag that you have assigned it. In the list include the manufacturer's name and model number and the important characteristics that describe each device.

Operation Procedure: Have your instructor ensure that the controller is correctly tuned. Mark in pen on the chart all your changes to the control loop as you make them. Set the loop in automatic (closed loop) operation at 50% of full tank level with a moderate amount of liquid flowing out of the tank (the manual valve position at 50% open). If, at any time, it looks as if the tank might overflow, quickly close the valve on the inlet line to the tank. Obtain a record of the variation in the level and valve position as the level settles at or near 50%. Change the level set point to 75% and obtain another record of the level and valve stem changes until they have settled at or near 75%.

Switch the controller to manual (open loop) operation and adjust the inlet valve manually to 5% open. Open the outlet valve to 50% open. Record the value of the level and valve position as they change and the level settles out at 0%. Retain the controller in manual (open loop) with the inlet valve at 5% open and close the flow of liquid to the outlet. Record the value of the level as it rises and, at 25% of full tank level, reopen the outlet valve to the previous 50% position. Manually adjust the inlet valve opening to 80% and record the time it takes to fill the tank to 65% full, and then switch the controller back to automatic (closed loop) mode at a level set point of 35%. Let it settle and then close the tank outlet valve and record the value of level and valve position as they settle. Shut down this control system by shutting off the water, air, and electricity.

---

*See Woodford, A. T. 1987. *TANKSIM.* ATW Software.

Conclusions: What do you do to the control loop when you switch from automatic to manual mode? Where does the change occur in the loop? What signal in Figure 1.1 do you adjust when the loop is in manual mode? When the loop is in automatic (closed loop) control, how do you disturb or upset the loop? How does the controller affect the loop in manual and in auto modes?

REFERENCES

1. Webb, John W. and Reis, Ronald A. 1995. *Programmable Logic Controllers, Principles and Applications.* 3rd ed. Upper Saddle River, NJ: Prentice Hall.

2

MEASURING SENSORS

OBJECTIVES

When you complete this chapter you will be able to:

- Prepare a static calibration certificate and graph for a measuring sensor
- Apply thermocouples for temperature measurement
- Apply RTDs (Resistance Temperature Detectors) for temperature measurement
- Specify pressure and differential pressure transmitters
- Specify an orifice plate for flow measurement
- Select a flowmeter
- Select a measuring sensor for liquid level or solids level measurement
- Describe force and weight load cells
- Measure humidity using a lithium chloride element and a sling psychrometer
- Measure electrolytic conductivity using a conductivity cell
- Describe applications of measuring sensors for electric power variables
- Prepare a dynamic calibration and frequency response of a measuring sensor

Recalling Figure 1.1, the purpose of a measuring sensor is to measure accurately the value of the control loop process variable. Usually the sensor is required to convert the measurement into a signal that can be readily communicated to the control loop controller block. Given current industrial applications, this signal will usually be electrical, although pneumatic signals are still very common, and optical transmission signals are just beginning to be used.

The sensor consists of a primary sensing element, which is inserted into the process fluid or material or attached to the process equipment, and a communications element. Examples of primary sensing elements include a thermocouple, which is often mounted in a thermowell and inserted into the process piping, or a vibration sensor, which is mounted on the bearings of heavy equipment. The communications component, located nearby or packaged with the primary sensing element, is usually an electronic amplifier or microprocessor that sends standard analog or digital signals corresponding to the primary sensing element's signal to the controller block. The sensor includes provision for connecting the communication line (electric, pneumatic, or optical) and the operating power (air supply, electricity, or solar). Power supplies for sensors are described in section 4.5.

Sensors are becoming much more sophisticated. A stylish term used to describe recent devices incorporating microprocessors is "smart" as in a *smart transmitter.* Some even include one microprocessor for communication purposes and another microprocessor for data acquisition. Digital communication as well as analog transmission is becoming common. Via the communication line, adjustments of span and zero in software, and diagnostic reporting of faults are now available. Several of the standard process sensors are described in this chapter. However, many of the less common and more complex sensors, including vibration, torque, and most chemical analysis sensors have not been included.*

2.1 CALIBRATION

Calibration usually consists of a check on the steady-state accuracy of a sensor. This is the most common type of calibration. However, the other blocks in the control loop (see Figure 1.1) may also be calibrated at the same time. For example, if a level transmitter is sending a signal to a recorder in the control room from a tank several hundred feet away on the process, the technician may set up calibration equipment near the transmitter. In the control room, the technician may have an associate with whom he or she communicates by means of a walkie-talkie. The technician will then proceed to inject known values of pressure (corresponding to level) into the transmitter, and the technician or the associate will then record: (a) the known injected values, (b) the readout values in the control room, and (c) the intermediate values generated by the transmitter signal. If the errors in the transmitter's intermediate signal are very small, then the technician may only make calibration changes to ensure that the control room readout values are accurate. If the loop is part of a computer system, the calibration results may be recorded automatically by the computer for a historical record of the loop device's calibration stability.

In many cases it is becoming desirable to establish the value of the dynamic calibration of the components in a loop (see section 2.11). It is usually not possible to adjust the dynamic calibration, and the user must take this into consideration. However, it may be desirable to know what this value is so that the control loop can be designed to perform in the best possible way.

Steady-State Calibration

The purpose of steady-state calibration is to provide confidence to the user that the sensor is stable, and maintaining its accuracy over its operating range as time passes. At least once a year, each sensor should be compared to a standard sensor that has

*Most of these sensors are described in Liptak and Venczel (1982) and Considine (1985). See the chapter references.

a calibration traceable to a national laboratory. The smaller the drift in the calibration, the more stable the device.

Steady-state calibration is the act of certifying that a sensor, over its operating range, has a specified set of errors found from comparing it to a designated standard device. The certificate, shown in Figure 2.1, will consist of a table of "as-found" values relating to the values of the standard device at approximately the 0%, 25%, 50%,

INSTRUMENT CALIBRATION CERTIFICATE

Identification: HPT483F601 — Input Range: 3–15 psig — Output Range: 4–20 ma DC

Date: May 18, 1996 — Mfr: FOXBORO — Model: E91

A Standard Input Signal (psig)	B Percent Input Signal (%)	C Desired Output Signal (ma DC)	D Actual Output Signal (ma DC)	E Actual Output Error (ma DC)	F Percent Output Error (%)	G Percent Output Signal (%)
3.03	0.25	4.04	4.03	–0.01	–0.0625	0.1875
6.05	25.417	8.067	8.05	–0.017	–0.106	25.31
9.02	50.167	12.027	12.08	+0.053	+0.331	50.5
12.06	75.5	16.08	15.98	–0.10	–0.625	74.88
15.00	100.0	20.00	20.05	+0.05	+0.313	100.31
11.95	74.583	15.933	15.92	–0.013	–0.081	74.50
8.97	49.75	11.96	12.00	+0.04	+0.25	50.00
6.03	25.25	8.04	7.95	–0.09	–0.563	24.69
3.01	0.0833	4.013	4.01	–0.003	–0.019	0.0625

Remarks: As Calibrated

Signature: Roy Fraser

FIGURE 2.1
Instrument calibration certificate

75%, and 100% of range of the sensor with calibrating signal rising from 0% to 100% and then with it falling from 100% to 0%. If the error is excessive at any of these values, then the sensor will be adjusted until the error is acceptable and another set of "as-calibrated" values will be recorded. A calibration graph (the upper graph on Figure 2.2) of the standard device (Percent Input Signal) values on the horizontal axis

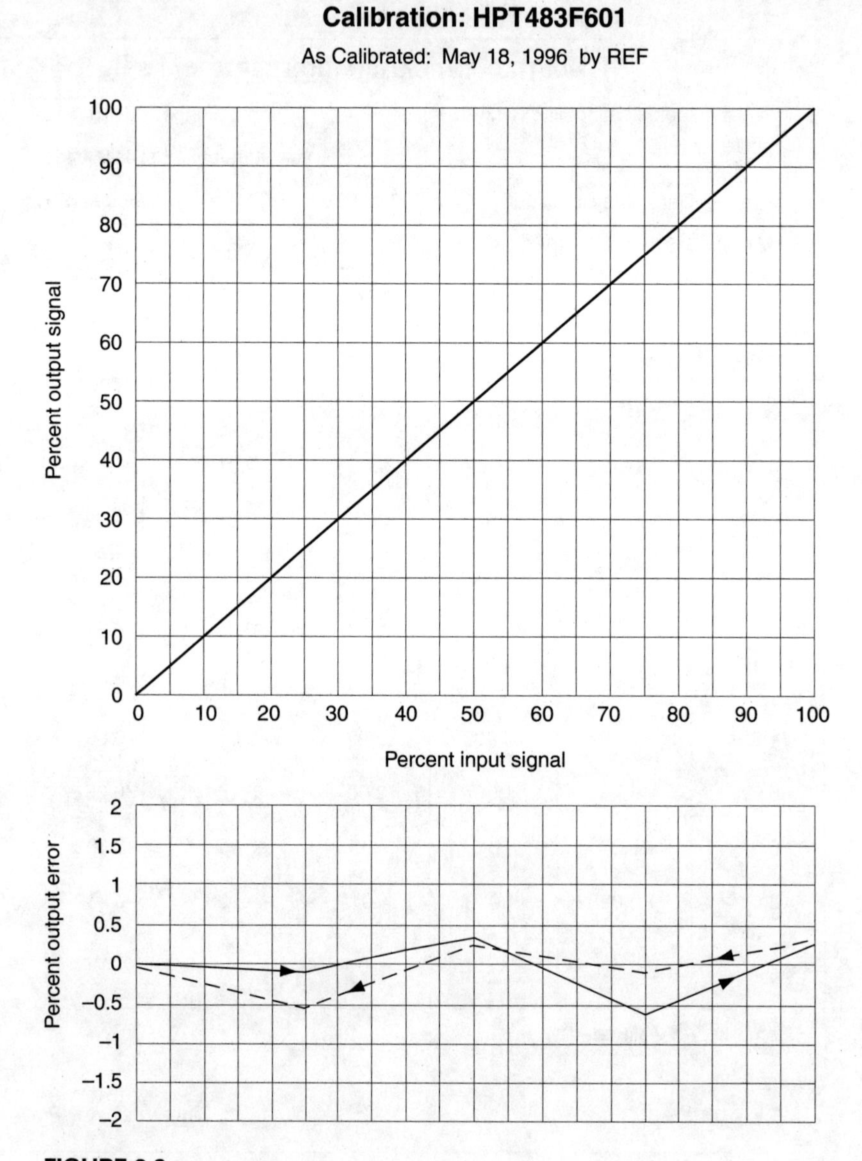

FIGURE 2.2
Calibration and deviation graphs

and the sensor (Percent Output Signal) values on the vertical axis will also be prepared for the certificate. A deviation graph (the lower graph on Figure 2.2), showing the hysteresis or deviation cycle (Percent Output Error), amplifies the error so that it is very apparent where and by how much the errors are. A typical certificate is shown for "as-calibrated" results in Figure 2.1 with associated graphs in Figure 2.2. On the deviation graph a solid line shows rising values (from minimum to maximum range) during calibration, and a dotted line shows falling values.

Notice that the input range for the pneumatic-to-current converter in Figure 2.1 is 3 to 15 psig and the output range is 4 to 20 ma DC. Column C, the desired output signal, is calculated from column B for an ideal (no error) converter. The actual output error, column E, is the difference between column C and column D, and the percent output error, column F, is the actual output error, column E, divided by the output span (16 ma), all multiplied by 100%.

Each column (A, B, C, D, E, F, G) of Figure 2.1 may be calculated as follows:

A = as measured on the designated standard reference device

$$B = \frac{(A - Z_i) \times 100\%}{S_i} \qquad \text{e.g.,} \quad \frac{(A - 3) \times 100\%}{12}$$

$$C = \frac{Z_o + (B \times S_o)}{100\%} \qquad \text{e.g.,} \quad \frac{4 + (B \times 16)}{100\%}$$

D = as measured on the device being calibrated

$$E = D - C$$

$$F = \frac{E \times 100\%}{S_o}$$

$$G = \frac{(D - Z_o) \times 100\%}{S_o}$$

where: Z_i = zero value of the input range, e.g., 3 psig
S_i = span value of the input range, e.g., 12 psig
Z_o = zero value of the output range, e.g., 4 ma
S_o = span value of the output range, e.g., 16 ma

2.2 PROCESS VARIABLES AND SELECTION OF SENSORS

The most common process variables are identified in the Instrument Society of America standard ISA-S5.1 in Appendix A. The major ones are temperature, pressure, flow, and level. For each type of process variable there are usually several types of sensors made by several different manufacturers. The process control designer must select the type of sensor that will cause the least problems for his particular application. He must discover the potential problems and weigh them against one another in order to make an effective choice.*

When selecting and specifying a measuring sensor, there are several major characteristics which should be considered. The following list, with examples, includes most of these characteristics:

- Process variable and its engineering units, e.g., temperature in degrees Celsius.
- Span adjustability, the ability to adjust the range of values of the process variable that correspond to the output signal range, e.g., the output range of 4–20 ma has a span of 16 ma that may be set to correspond to a minimum input span of 5°C or up to a maximum input span of 250°C.
- Zero adjustability, suppression or elevation of the process variable zero relative to the output zero signal, e.g., the process variable value of zero may be suppressed to be below the output zero of 4 ma or elevated so that it corresponds to an output greater than the output zero of 4 ma.
- Precision, the calibration error to which the device may be readily adjusted, e.g., ± 0.25°C in the linear relation of the output signal to the measured process variable.
- Resolution, the minimum, detectable variation in the output signal caused by a corresponding change in the process variable, e.g., 0.05°C.
- Stability, maximum expected change in calibration error with time, e.g., 0.1°C per six months.
- Communication signals, e.g., analog: 4–20 ma DC or digital: HART protocol.
- Process connections, e.g., ¼″ NPT.
- Mounting, e.g., 2″ NPT pipe stand.
- Power supply, e.g., 13–30 V DC or 115 V 50–60 Hz.
- Environmental effects, e.g., error in degrees C per degree C of ambient temperature change.

Vendors of measuring sensors (such as ABB, Foxboro, Bailey Meter, Honeywell, and the like) will provide brochures for any standard device available from them. Phone the local representative or look them up on the Internet for information about any device for which you require additional details. When you require a measuring sensor, vendors will be pleased to quote to your specifications and advise you how to correctly apply their particular type of sensor.

---

*Liptak and Venczel (1982), Tippens (1978), and Considine (1985) explain the units and the physics that apply to each of the sensors described in this chapter. For example, Chapter 17 of Tippens (1978) on temperature explains the relationship of temperature to thermal energy and it explains the various temperature scales, including the absolute scales of Kelvin and Rankine.

2.3 TEMPERATURE

The different types of common temperature sensors are classified.* These common types of sensors include:

1. Mechanical Sensors:
   - Bimetallic Devices
   - Liquid or Gas Filled Thermal Systems
2. Electrical Sensors:
   - Thermocouples
   - Resistance Temperature Detectors (RTDs)
   - Semiconductors (including thermistors)
   - Radiation Pyrometers
3. Optical Sensors (not yet common)

The mechanical sensors are not as convenient as the electrical ones for groups of variables greater than six or for distant data acquisition (more than 25 meters). Therefore, the sensors described in this section will include only the electrical (thermocouples and RTDs) sensors.†

Thermocouples

Since 1821, when German scientist Thomas Seebeck took two dissimilar metal wires and connected them as shown in Figure 2.3, thermocouples have been applied to temperature measurement. The measured voltage, E_t, is equal to the voltage $E_{ic1} - E_{ic2}$, where E_{ic1} is the voltage generated by the temperature T_1 from the hot iron-copper junction if it were connected to a reference iron-copper junction held at 0°C, and E_{ic2} is the voltage generated by the temperature T_2 from the cold iron-copper junction if it were connected separately to another reference junction at 0°C.

Tables exist (see Appendix F‡) showing temperatures with associated voltages for a variety of dual dissimilar metals, including the following:

+Metal	*–Metal*	*Type*	*Average mv/°C*	*Useful Range*
Iron	Constantan	J	0.054	0–800°C
Chromel	Alumel	K	0.0414	0–1200°C
Platinum	Plat-rhodium	R	0.009	500–1500°C
Copper	Constantan	T	0.045	–100–400°C

*For additional information about common sensors and their classifications, see Anderson (1980) and Liptak and Venczel (1982).

†For details on the mechanical sensors, see Kirk and Rimboi (1975), Anderson (1980), and Liptak and Venczel (1982). For details on semiconductors and radiation pyrometers, see Liptak and Venczel (1982) and Considine (1985).

‡Relevant tables are also reproduced in Liptak and Venczel (1982).

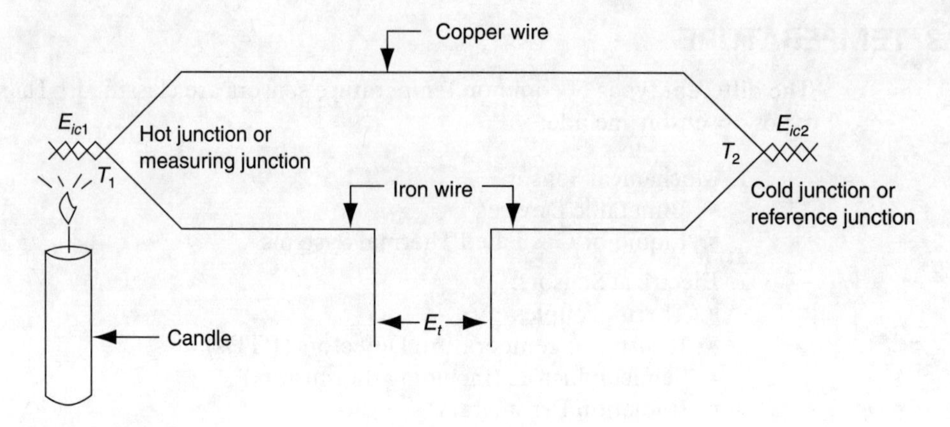

FIGURE 2.3
Thomas Seebeck's "thermocouple" experiment of 1821

Notes
Constantan is a mixture of 45% nickel and 55% copper.
Chromel is a mixture of 90% nickel and 10% chromium.
Alumel is a mixture of 94% nickel, 2.5% manganese, 2% aluminum, 1% silicon, and 0.5% iron.
Platinum-rhodium is a mixture of 87% platinum and 13% rhodium.

The average output is a very low voltage, in millivolts per degree Celsius of temperature difference, between the measuring junction and the reference junction.

The Law of Intermediate Metals applies to thermocouple circuits that have the usual two dissimilar metals at the measuring junction, but which introduce a third metal near the device that measures the millivoltage being produced. As shown in Figure 2.4, if there is no difference in temperature between T_2 and T_3, then the junction at T_2, between chromel and copper, and the junction at T_3, between alumel and copper, will not introduce any net voltage to affect the voltage, E_t, generated by the measuring chromel–alumel junction at T_1 relative to the reference chromel–alumel junction. In effect the reference junction of chromel–alumel is at the temperature $T_2 = T_3$. Thus if T_2 and T_3 are at terminals in the measuring instrument, and we measure their temperature (which is not expected to differ from one terminal to the other by more than 0.1°C), then we can find their reference junction millivolts (relative to

FIGURE 2.4
Intermediate metals

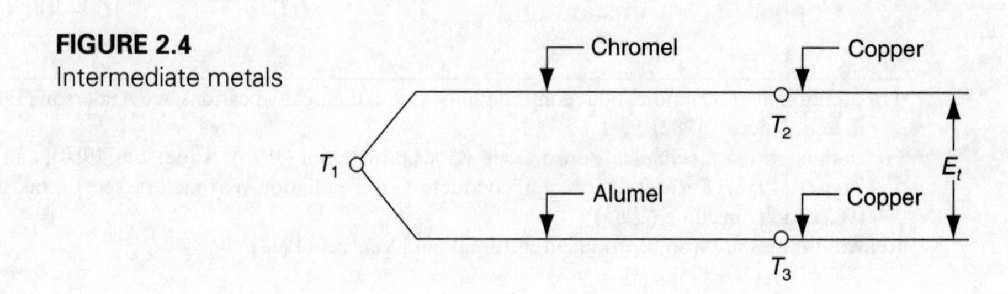

0°C) in the tables. Since we will have measured accurately the chromel–alumel millivoltage, E_t, then the hot junction millivolts (relative to 0°C) will be the reference junction millivolts plus E_t, and via the tables we will have the temperature at T_1.

EXAMPLE 2.1 **Thermocouple: Temperature–Millivolt Relationship**

As shown in Figure 2.4, a chromel–alumel thermocouple measures an unknown temperature in a hot-water tank. The thermocouple is mounted in a thermowell which is immersed in the water of the tank. On a nearby control panel, a millivoltmeter recorder is connected by (chromel–alumel) thermocouple extension wire to the thermocouple. A thermometer measures the (reference junction) temperature (T_2 and T_3) at the terminals inside the recorder at 21.3°C. The millivoltmeter is recording a value of 3.111 mV for E_t. What is the temperature of the thermocouple (and the hot water in the tank)?

Recorder reference junction millivolts: From tables (in Appendix F for type K) at 20°C is 0.798 mV; at 25°C is 1.000 mV; at 21.3°C is 0.798 + (21.3 – 20.0) × (1.000 – 0.798) / (25 – 20) = 0.851 mV.

The thermocouple voltage, when the reference junction is at 0°C (rather than 21.3°C), is 3.111 + 0.851 = 3.962 mV. From the tables at 3.888 mV, the temperature is 95°C; and at 4.095 mV the temperature is 100°C. The temperature at 3.962 mV is 95 + (3.962 – 3.888) × (100 – 95) / (4.095 – 3.888) = 96.79°C.

An additional Law of Thermoelectricity applies, namely the Law of Intermediate Temperatures, which states that the voltage, E_a, generated by thermocouple A (see Figure 2.5) will have the same value as the sum of the voltages, E_b and E_c, from two similar thermocouples, B and C, connected thermally in series, but electrically isolated. This will be so if the hot junction of A is at the same temperature, T_1, as the hot junction of B and the reference junction of B, is at the same temperature, T_3, as the hot junction of C and the reference junction of C is at the same temperature, T_2, as the reference junction of A. This law will allow the temperature of a reference junction to be measured at room temperature rather than at 0°C, and a simple numeric correction gives us the corrected millivolts, and from the tables we arrive at the hot junction temperature.

Usually the reference junction temperature is measured by means of a resistance temperature detector (RTD) located at the terminals of the measuring millivoltmeter, where the thermocouple extension wire terminates. The circuit containing the RTD generates a millivoltage (equivalent to that of a thermocouple with a hot junction at room temperature and a reference junction at ice-water temperature) that is added to the measured hot junction millivolts to compensate for the reference junction not being at ice-water temperature.

Thermocouples are often insulated electrically with ceramic material and sheathed in stainless steel (see Figures 2.6 and 2.7) and they are often protected by thermowells which project into the process fluid and allow maintenance personnel to replace them while their process continues operating. Sheathed thermocouples in

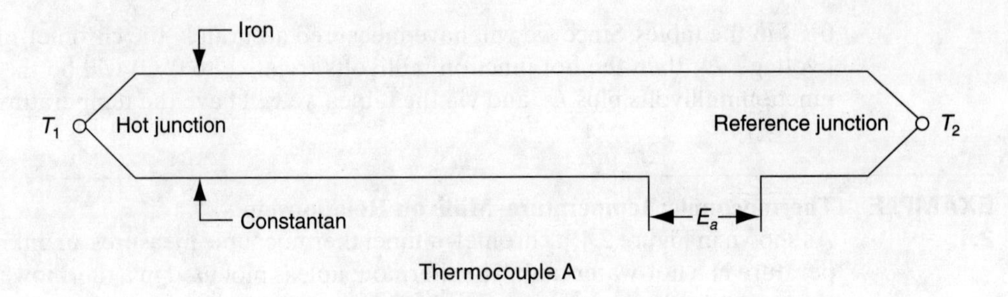

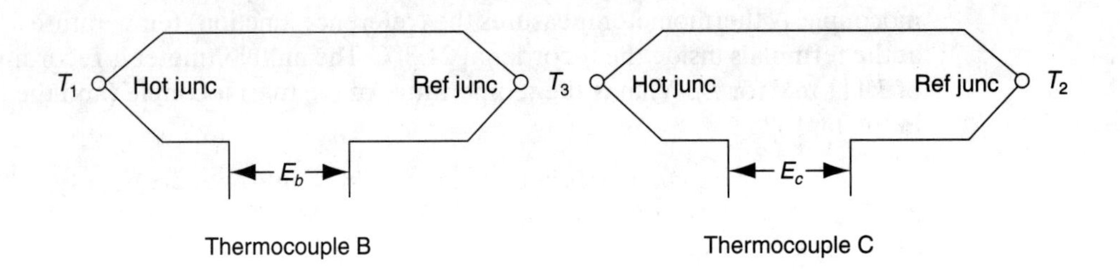

$$E_a = E_b + E_c$$

FIGURE 2.5
Intermediate temperatures

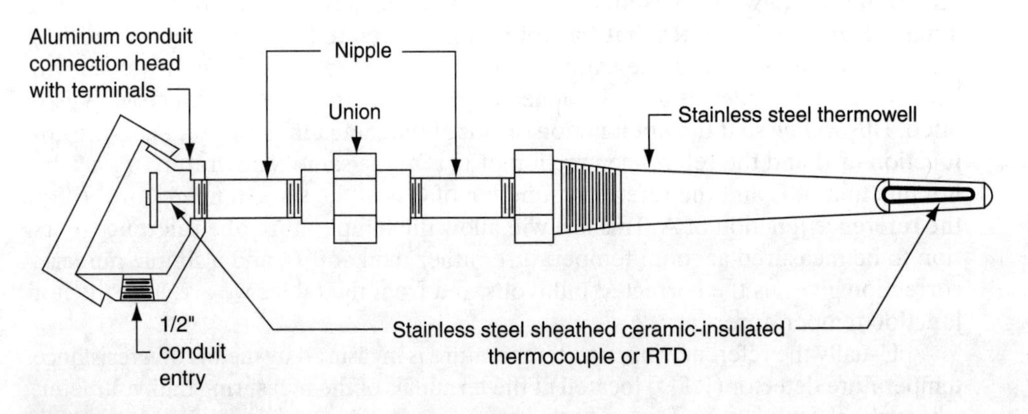

FIGURE 2.6
Temperature sensor (time constant ~100 seconds in water)

wells have long time constants—100 seconds in water is common. Very thin (1/16″) sheathed and grounded thermocouples without a well may have a fast time constant of less than one second in water.

Thermocouple errors may be introduced by impurities in the dissimilar metals, by inaccuracies in the measurement of the reference junction temperature, by

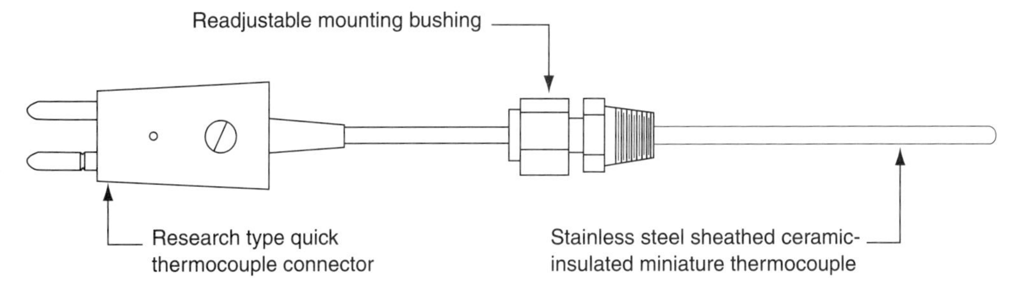

FIGURE 2.7
Temperature sensor (grounded thermocouple with no thermowell; time constant ~0.2 seconds in water)

errors in the measuring millivoltmeter, and interference from radiating electric and magnetic circuits. If the interference is negligible, then the other errors may be as low as ±0.5°C for copper–constantan over the range –50°C to 100°C, or ±5°C for platinum–platinum/rhodium over the range 500°C to 1500°C.*

Many manufacturers (Fisher-Rosemount, ABB, Honeywell, and so on) make thermocouple transmitters that include reference junction compensation and convert the thermocouple millivolts to a 4–20 ma signal, corresponding to a desired temperature span, for transmission to a computer control system. These transmitters may even be mounted in the conduit-connection head of the thermocouple. Often the thermocouple error itself greatly exceeds that of the transmitter so that the transmitter error has no significant effect.

RTDs (Resistance Temperature Detectors)

In 1871 Sir William Siemens first described the use of platinum resistors for temperature measurement. They are now the international standard for temperature measurements between 13.81°K (triple point of hydrogen) and 903.90°K (freezing point of antimony). The most common RTDs are made from platinum. For some very simple applications, however, copper is used, and for some other applications nickel is used. The stability and repeatability of platinum makes it the standard of the research laboratory.†

Resistance Externally powered or self-powered resistance transmitters (or transducers) are usually used with resistance as a secondary variable to infer a primary variable. For example, an RTD senses temperature (the primary variable) and converts it to resistance (the secondary variable) and the resistance transmitter converts this resistance to a 4–20 ma DC signal for transmission to a computer system. Another example is the use of a resistance transmitter to convert the resistance of the position sensing potentiometer on a damper or a control valve to 4–20 ma DC for

*For error estimates, see Liptak and Venczel (1982).

†For a description of the requirements for fabricating industrial RTDs, see Liptak and Venczel (1982). This handbook also shows the assembly of an RTD in a thermowell.

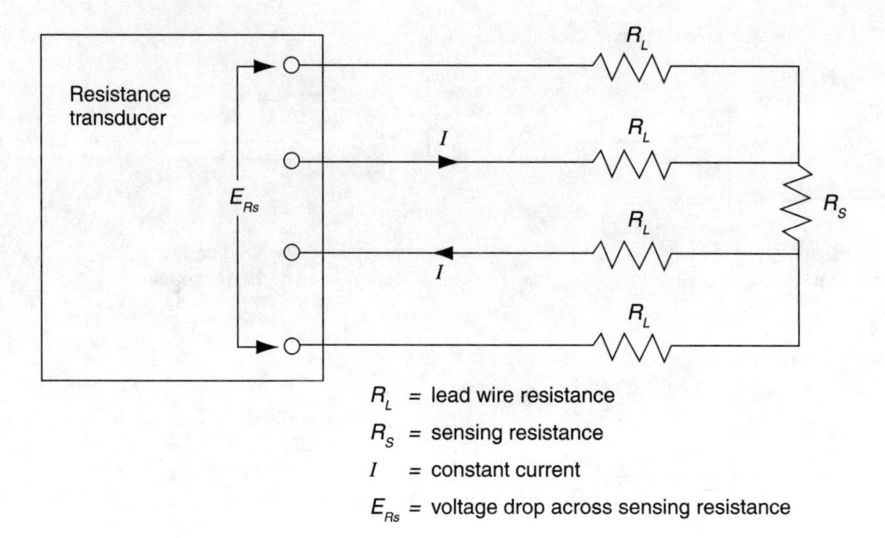

FIGURE 2.8
Four-wire measurement of resistance

transmission to the computer. If the sensing resistance is nonlinear with respect to the primary variable, then it may be best for calibration purposes to keep the transmitter output linear with respect to resistance and use tables in the computer to establish the value of the primary variable. As shown in Figure 2.8, a four-wire measurement of resistance will give the most accurate results. For example, if the lead wire resistance will affect the measurement of the sensing resistor significantly, then four-wire measurement becomes desirable. The resistance transmitter sends a constant current, I, through the sensing resistance, R_s, and only the voltage drop across R_s is measured; the voltage drop across the current carrying lead wires is ignored. Many RTDs are supplied with four leads that connect to the platinum element right at the tip of the probe. Since the voltage, $E_{Rs} = I \times R_s$, and I is constant, then E_{Rs} is directly proportional to the resistance, R_s. The input impedance for the measurement of E_{Rs} is extremely high, so that the current flowing in the leads to the voltage measurement point for E_{Rs} is much less than I, and the voltage drop in those leads is much less than the voltage drop in the leads where I flows.

Many manufacturers (e.g., Fisher-Rosemount, Foxboro) make resistance transmitters that convert the RTD resistance to a 4–20 ma signal, corresponding to a linearized temperature span, with an overall temperature error less than 0.2% of span for spans down to 5°F.

EXAMPLE 2.2 **RTD: Resistance–Temperature Relationship**

A four-wire platinum RTD based on DIN curve 43760 is used to measure the temperature of a bearing on a heavy machine. It is mounted in a cavity of the bearing. The ohmmeter shows a value of 118.23 Ω for the RTD. What is the temperature of the RTD and therefore the temperature of the bearing?

From the table in Appendix F for Platinum RTD 100 Ω @ 0°C, DIN Curve 43760 at 117.47 Ω, the temperature is 45°C; and at 119.40 Ω the temperature is 50°C. The temperature at 118.23 Ω is 45 + (118.23 – 117.47) × (50 – 45) / (119.40 – 117.47) = 46.97°C.

EXAMPLE 2.3 **RTD Specification**

Application: Monitor the temperature of the end bearings of a large, rotating mill over the range of 0°C to 200°C; transmit a 4–20 ma DC analog signal corresponding to the required temperature range to a central computer system.

Sensor: 6″ long, ¼″ O.D. ceramic-insulated, stainless steel sheathed, platinum RTD following DIN curve 43760, 9–68 with 100 Ω at 0°C.

Transmitter: Calibrated for 100 Ω RTD (curve DIN 43760) having a range of 0–200°C with 4–20 ma DC output.

Mounting: RTD to project from the transmitter with ½″ NPT mounting bushing to enter threaded bearing hole.

2.4 PRESSURE

The confusion of pounds-force and pounds-mass are generally overcome by using metric (SI) units: newtons (N) for force and weight, pascals (Pa) (newtons per square meter) and kilopascals (kPa) for pressure, and kilograms (kg) for mass. However, it is still desirable to understand units of pounds per square inch, inches of water, and inches of mercury vacuum, and to understand the effects of changes in atmospheric pressure on gauge pressure.*

Gauge pressure uses atmospheric pressure as a reference or zero point; on the absolute scale this reference (14.7 psia or 101.3 kPa absolute) shifts a small amount from time to time due to the weather, perhaps as much as ±7 kPa or ±1 psi.

Unit pressure is defined as unit force exerted on a unit area. The formula for pressure is

$$P\ (\text{pressure}) = \frac{F\ (\text{force})}{A\ (\text{area})}$$

where: P is in pascals or pounds per square inch
F is in newtons or pounds force
A is in square meters or square inches

At the earth's surface, the force of gravity causes 1 kg of mass to weigh approximately 9.8 N.

---

*For additional information about the important features of pressure, especially the concepts of absolute, gauge, and vacuum pressures, see Kirk and Rimboi (1975), Anderson (1980), O'Higgins (1966), and Tippens (1978).

Since:	F (force)	=	m (mass)	×	a (acceleration)
Then:	1 N	=	1 kg	×	1 meter/second/second
	9.8 N	=	1 kg	×	9.8 m/s^2 (acceleration of gravity)

Actually a mass of 1 kg weighs 9.806650 N at the standard location (Paris). But the same kilogram weighs 9.81276 N in the Netherlands and 9.79966 N in Melbourne, Australia. Therefore, if weights are used to produce pressure, then gravity must be taken into account for precise calibration.

A tank that is a cube with 1 meter on each side holds 1000 kg of water, weighing approximately 9800 N. The pressure on the bottom of this tank is 9800 Pa or 9.8 kPa. Thus a 100 cm column of water has a pressure at its bottom of 9.8 kPa and 1 cm of water column has a pressure of 98 Pa. One inch of water has a pressure of 98 × 2.54 = 249 Pa and 4 in. of water has a pressure of approximately 1 kPa.

The common electronic pressure sensor for data acquisition uses a diaphragm with a special device such as a capacitor (see Figure 2.11 on page 28), or a vibrating wire, or a strain gauge to sense the pressure produced by the process. These devices generally measure differential pressure. If they measure the pressure drop of a fluid flowing in a pipe, then the high pressure side of the diaphragm is connected upstream on the pipe and the low pressure side is connected downstream. If they measure pressure at a point on a pipe relative to atmospheric pressure, then the low pressure side of the diaphragm is open to the atmosphere, and gauge pressure is measured. If the low pressure side of the diaphragm is a chamber evacuated to 0 psia, then absolute pressure is measured.

Installation of Pressure Sensors

Correct installation of pressure measuring devices, especially differential pressure measuring devices, is dependent on the type of fluid being measured. The tubing that connects the sensor to the process is called an impulse line, because it contains the process fluid, and it transmits pressure impulses to the sensor. When differential pressure is measured, there are two impulse lines connected to the sensor. For liquids it is usually desirable to keep the impulse lines connecting the sensor to the fluid filled with the process liquid, so that on shutdown and startup of the process the impulse lines will remain full of liquid. This means that the impulse lines for process liquids should slope down from their process connections toward the pressure sensor as shown in Figure 2.9. Therefore, once the differential pressure sensor has had all the gas bled out of both impulse lines, and the impulse lines plus the sensor chambers are full of process liquid, they should remain full of liquid. Unfortunately they may also receive some sediment over time.

For sensors measuring process gases, except steam, correct installation usually demands that liquids not be allowed to accumulate in the impulse lines. This means that the impulse lines for process gases should slope up from their process connections toward the pressure sensor as shown in Figure 2.10. Thus, once the sensor has had all the air bled out of both impulse lines, they should remain full of process gas.

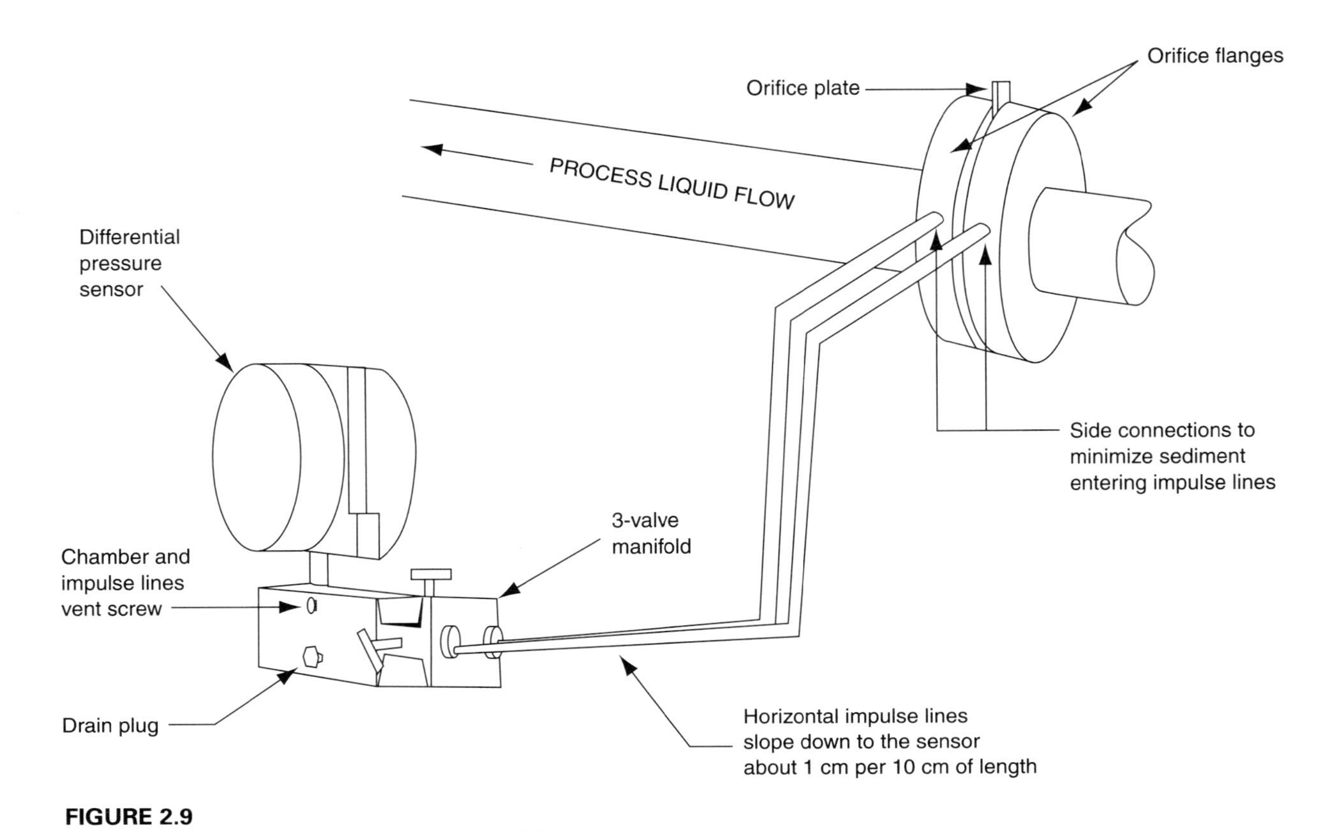

FIGURE 2.9
Impulse lines and sensor chambers for liquids

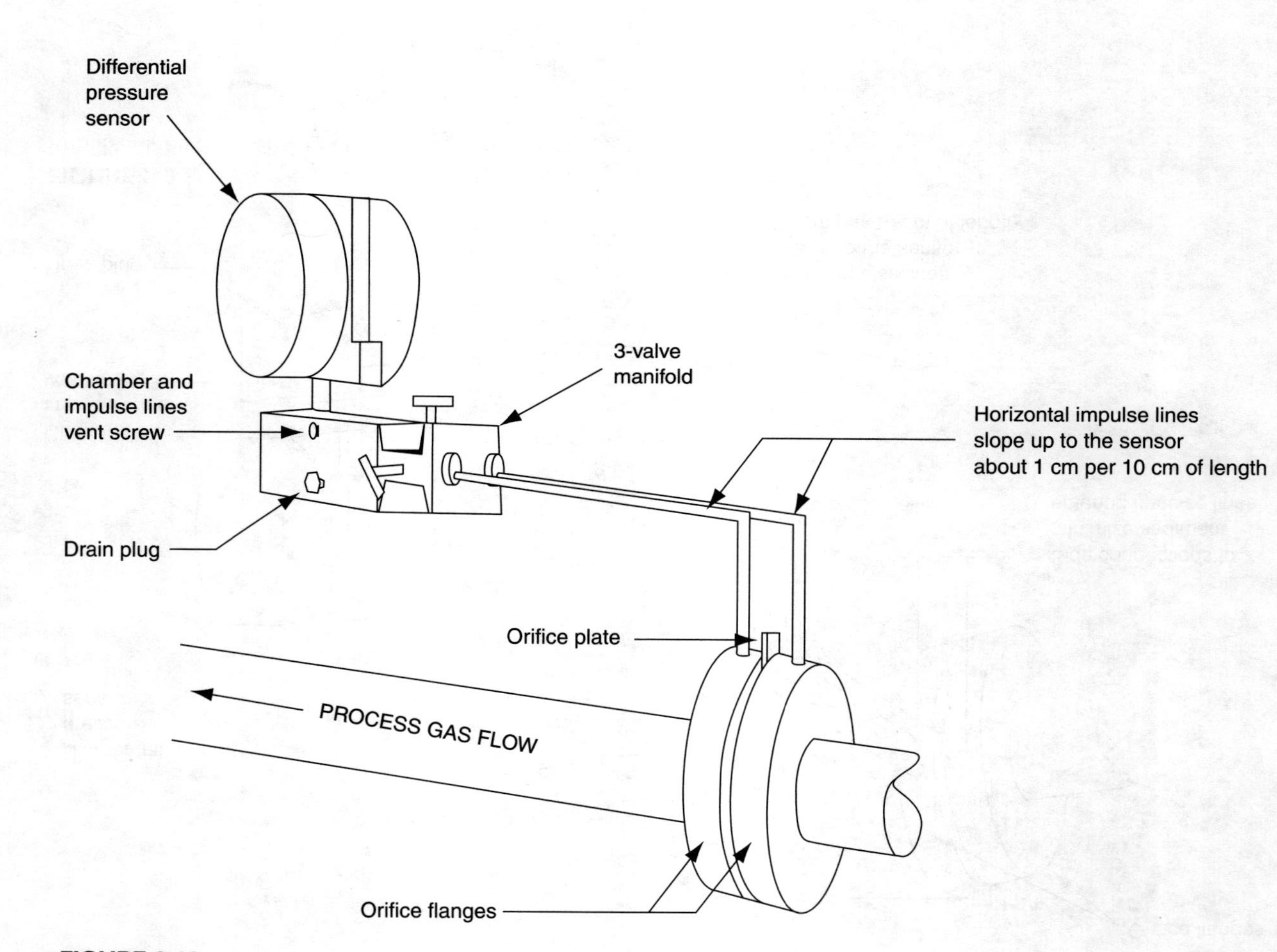

FIGURE 2.10
Impulse lines and sensor chambers for gases

For sensors measuring steam, correct installation requires steam condensate pots on each impulse line. The pot should be very close to the point where the impulse line connects to the process so that steam will continually condense in the pot and the water run back into the process. The measurement is then made with impulse lines as if for a liquid measurement, using the liquid water in the condensate pot to transmit the pressure via a liquid water-filled impulse line down to the sensor.

These devices are usually installed with a valve manifold that will allow them to be disconnected from the process at any time for calibration or maintenance purposes. For line pressure measurement, this is a single valve; and for differential pressure measurement, this is a combination of two isolating valves and an equalizing valve for zero calibration. The sensors also usually include gas venting and liquid drain screws in their body on both the high and low pressure sides. Calibration of these devices may be done at their field location using a transfer standard, such as a differential pressure transmitter referenced to a shop standard (usually a dead-weight tester).*

Capacitance Diaphragm Pressure Sensors

As shown in Figure 2.11, a differential pressure is transferred through a noncompressible fill liquid to slightly displace the measuring diaphragm (moving electrode) of a capacitance pressure sensor. The measuring diaphragm acts as the common capacitor plate between two fixed capacitor plates. The displacement increases the distance from the movable plate to one of the fixed plates and decreases the distance to the other fixed plate, resulting in a change in capacitance that is a precise function of the differential pressure. The change in capacitance is detected by electronic circuitry and transmitted to the data acquisition system. The electronic circuitry includes range adjustments with a 10:1 span turndown ratio and a zero elevation and suppression from +90% to –90% of the maximum range limit. The transmission signal can be 4–20 ma DC or digital over wires or via fiber optic line. The estimated error is less than 0.25% of the calibrated span and the stability is quoted as less than 0.25% of maximum range over six months. The device that receives the 4–20 ma signal usually provides the power supply for the differential pressure transmitter over the same two wires that the 4–20 ma signal uses.

Other manufacturers make a similar unit with an adjustable range having a span turndown ratio of 6:1 and a zero suppression adjustable up to 500% of span plus a zero elevation adjustable up to 600% of the span.

Strain Gauge Diaphragm Pressure Sensors

The bonded strain gauge diaphragm pressure sensor is one of the most common types of pressure transducer. Each of four arms of a wheatstone bridge are made up from strain resistance wire bonded to the diaphragm. Resistors are added to the circuit to

*For additional information on dead-weight tester, see Liptak and Venczel (1982).

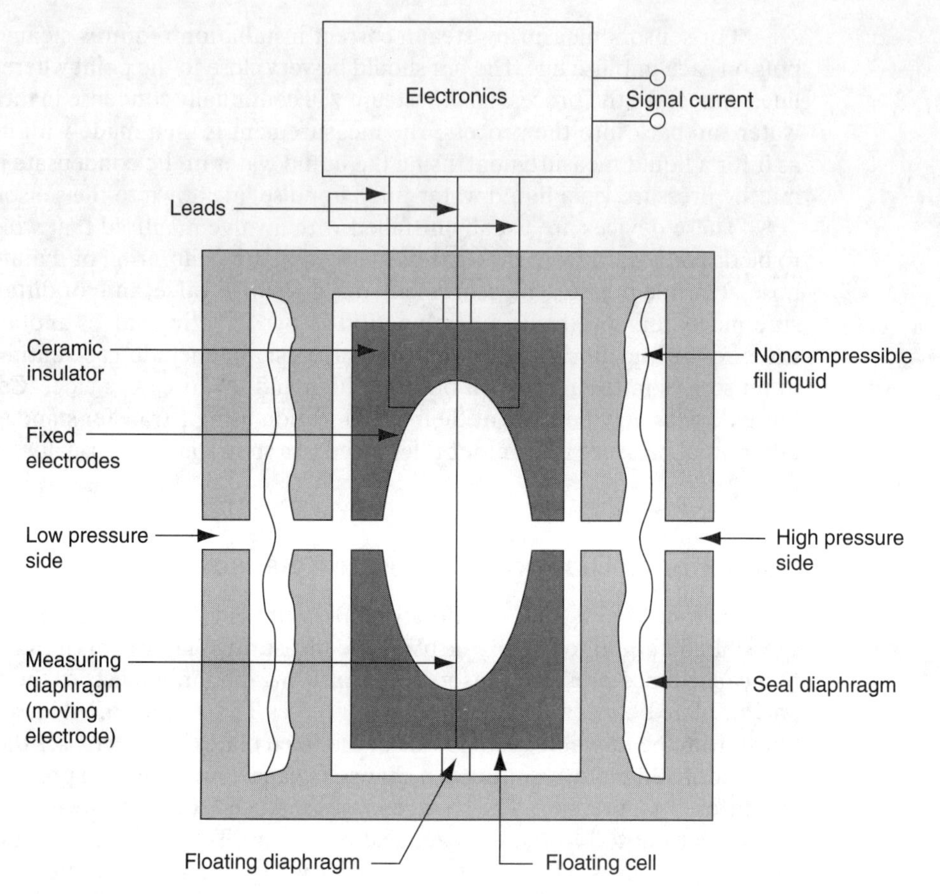

FIGURE 2.11
Capacitance diaphragm pressure sensor

provide temperature compensation and calibration adjustment. If a high level output is required, then an amplifier is provided in the housing. These units may be 10 cm long and 2.5 cm in diameter.

These units are available in fixed pressure ranges from 0–5 psia (pounds per square inch absolute), psig (pounds per square inch gauge), psid (pounds per square inch differential), and psis (pounds per square inch sealed with a reference pressure of 14.696 psia) to 0–10,000 psia, psig, psid, and psis. Each range is about 25% to 50% greater than the previous lower range.

The wheatstone bridge is usually excited with 10 V DC and a low level output at full scale of between 2 and 3 mV per volt of excitation is commonly achieved. When an amplifier is included to provide a high level output, then full-range outputs of 0–5 V DC or 4–20 ma DC are available; however, an excitation voltage of 28 V DC is required to power the amplifier as well as the wheatstone bridge. Errors for these strain gauge units are between 0.5% and 1% of full scale. Maximum overrange without damage is about 150% of full scale.

Very small (0.080 in. diameter) strain gauge pressure sensors with frequency responses up to 25 kHz are available for engine pressure measurements.

Vibrating Wire Diaphragm Pressure Sensors

This type of pressure sensor is made only by the Foxboro Company. A vibrating wire is stretched by the force generated from the process differential pressure applied to a diaphragm. The frequency of vibration is related to the tension in the wire. Electronics convert the frequency to a 4–20 ma DC signal. Thus the frequency is sensed and calibrated against the differential pressure applied to the diaphragm. The electronics include an adjustable span with a turndown ratio of 6:1 and a maximum zero suppression and elevation of 150% of span on condition that it is less than the upper-range limit. Errors are quoted as less than 0.2% of calibrated span. Units are available with a range as low as 1 kPa and other units have ranges up to 100 MPa (megapascals).

Quartz Bourdon Helix Pressure Sensors

Extremely accurate (error less than 0.04% full scale) pressure measurement is possible with the quartz Bourdon sensor. Many standard ranges from 0–1 psi to 0–2500 psi are available. Gauge, differential, and absolute pressure measurements are available. Resolution is 0.001% of range. Up till recently these sensors were relatively expensive compared to the other types, and they were retained for calibration purposes only. But now they are starting to be used for standard industrial applications requiring high precision measurement.

Force-balance quartz Bourdon sensors are constrained to function with very little mechanical motion. The tip of the Bourdon tube has a mirror, which tends to change its position as the pressure inside the Bourdon tube changes. Any error in the position of the mirror from its balanced position is detected optically and used to return the Bourdon tube to a balanced position using electromagnetic-force balancing coils in a servomechanism loop. The pressure inside the Bourdon tube is proportional to the constraining magnetic force and thus the electric current in the coils. Wear is reduced to a minimum, and the Bourdon tube is retained at its optimum mechanical dimensions. Standard analog outputs of 0–5 V DC and 4–20 ma DC and a variety of digital outputs are available.

2.5 FLOW

The process variable most commonly measured is flow rate. The variety of devices that are available for this measurement is immense. Flow of material (liquid, gas, or solid particles) is the most common flow measurement; however, flow rate of energy is also required. Flow rate of energy is power, and electric power measurement is very common. The integral of power, energy used, is also sometimes required, and is fairly easily measured as an electric variable (the residential watt-hour meter is an example). Flow of thermal (steam) energy, or hydraulic energy is much less commonly required and usually relies on material flow (steam or water) measurement plus temperature or pressure measurement. This section will concentrate on the most common devices for measuring and transmitting signals of material flow rates. The total material that passes through the flowmeter from an initial time until some arbitrary time can be calculated by integrating the flow rate in the computer system.

Liquid flow measurement is fairly easy to understand. For example, water coming out of a tap may be directed to a calibrated glass container. A timer is started when the water is at a low mark on the container and is stopped when the water reaches a high mark, showing the filling time for the volume between the high and low marks. The flow rate, in liters per second, is the quantity of water in liters between the two marks divided by the number of seconds that elapsed during the filling time. If we were using this flow rate to calibrate another flow rate meter in the line, then we would want the flow rate to remain as constant as possible during the filling period.

Gas flow measurement can be understood by the filling of a propane gas bottle. Here the empty bottle is placed on a weighing scale and the flow is started. When the scale is at a low mark or value, a timer is started. When the scale reaches a high value, the timer is stopped, and it shows the filling time. Therefore the flow rate, in kilograms of propane per second, is the weight of propane (found by subtracting the low value from the high value) and dividing by the number of seconds that elapsed during the filling time. Again, a constant flow rate is required during the filling time.

Solid flow rates are measured with conveyor belts that carry particles or lumps of the solid. On a short (2 m) length of the conveyor belt, the weight (in kilograms per meter of conveyer belt length) of solid material multiplied by the speed (in meters per second) of the belt will give the flow rate (in kilograms per second) of solid material flowing past the weighing point. These solid flowmeters are called *weighfeeders* and their associated electronics contain features for multiplying the weight sensor signal by the belt speed sensor signal and then comparing this with a set point to adjust the motor speed driving the belt (see Figure 2.12).

The most common fluid (liquid or gas) flowmeters with transmission capability are

- Orifice plate, or venturi, plus differential pressure transmitter
- Vortex shedding flowmeter
- Magnetic flowmeter
- Ultrasonic flowmeter
- Turbine flowmeter

Two common fluid flowmeters that seldom have transmission signal capability and, therefore, are not considered for process control are

- Positive displacement (residential gas and water meter)
- Rotameters

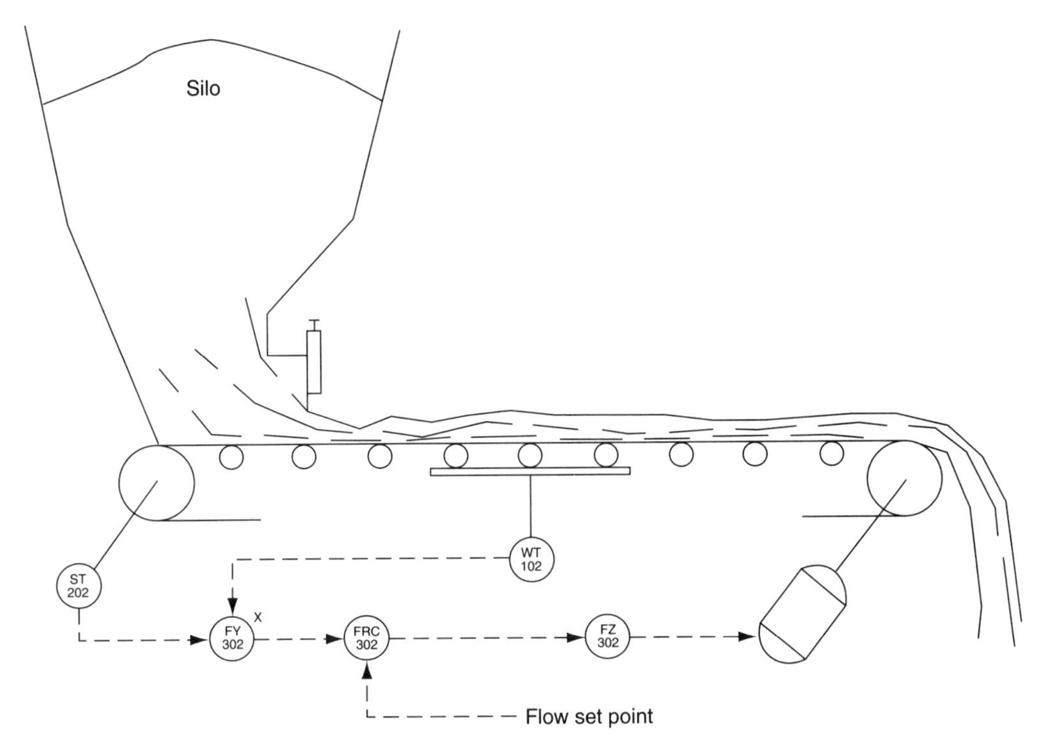

FIGURE 2.12
Flowmeter for solids

Only the magnetic and the ultrasonic flowmeters do not introduce restrictions into the fluid pipeline. All the other flowmeters interfere with the flow of fluid. With clean fluids this is usually no problem. With dirty fluids or slurries the wear on the instrument obstructions may quickly introduce excessive error to the measurement accuracy.

For best performance it is recommended that every flowmeter have a certain amount of straight pipe upstream and downstream from itself. For orifices, 10 to 20 pipe diameters upstream and 2 to 5 pipe diameters downstream are usually sufficient.*

Orifice Plate or Venturi

The orifice plate and venturi flowmeters introduce constrictions into the fluid flow path such that the flow velocity is suddenly increased and the pressure of the fluid is suddenly decreased.†

*For pipe with an unusual number of bends and disturbances, such as valves and pipe fittings, see section 2.12 in Liptak and Venczel (1982).

†For a detailed description, see Chapter 16 in Tippens (1978).

An orifice plate is shown in Figure 2.13 and a venturi in Figure 2.14. The advantage of the venturi over the orifice plate is that downstream of the constriction the venturi recovers up to 85% of the pressure drop, while the orifice may recover only 50% of the pressure drop.* The saving in pumping costs has to be balanced against the extra capital cost for the venturi over the orifice plate and flanges. The pressure drop across the orifice is measured with a differential pressure transmitter (see section 2.4). These flowmeters are appropriate for turbulent flows having Reynolds numbers greater than 10,000, which exists in more than 90% of all industrial flow measurement applications.

Reynolds number, R_D, is a dimensionless value related to fluid velocity and calculated as follows:

*These unrecovered pressure losses are described in section 2.23 of Liptak and Venczel (1982).

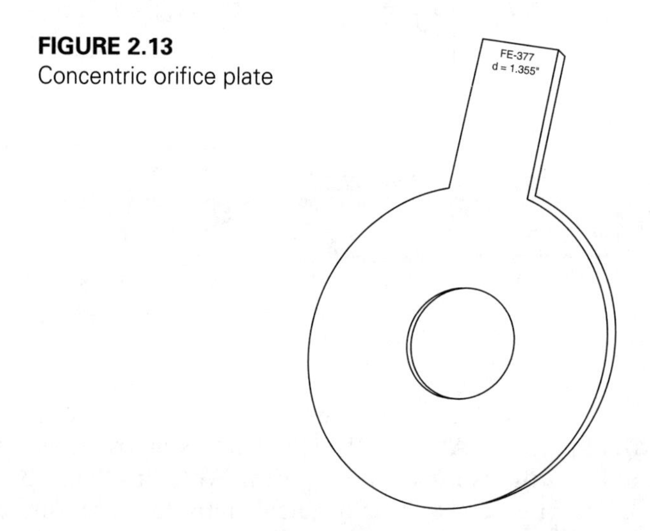

FIGURE 2.13
Concentric orifice plate

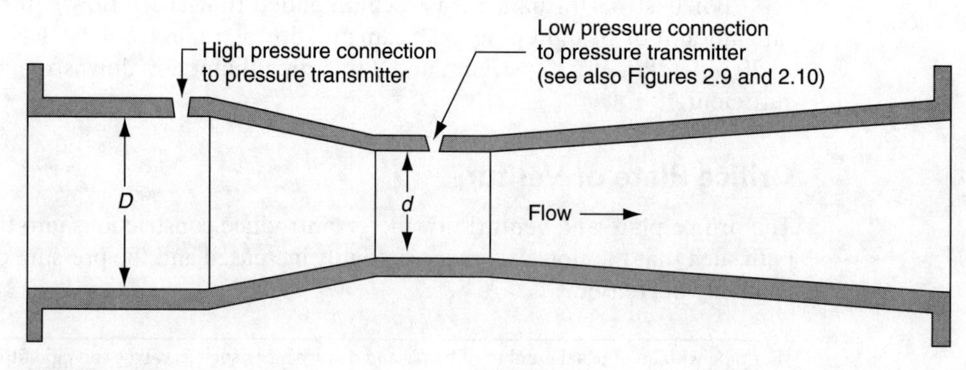

FIGURE 2.14
Venturi flowmeter

$$R_D = \frac{48 \times w}{\pi \times D \times \mu_1 \times g} = \frac{W}{235.6 \times D \times \mu_1 \times g}$$

where: R_D = Reynolds number based on pipe diameter
w = flow rate, lb/sec
π = 3.14159
D = pipe inside diameter, inches
μ_1 = inlet viscosity, lb(force) – sec/sq ft
g = gravity, ft/sec/sec (32.2)
W = flow rate, lb/hr

In the design of these head flowmeters (so named because flow is related to a loss of pressure head), the major assumption based on thousands of practical confirmations is that a coefficient of discharge, K, can be interpolated from tables,* such as Table 2.1, that relate K to the pipe Reynolds number and ß (beta) ratio (ratio of orifice diameter or venturi throat diameter, d, to pipe inside diameter, D). Even though Reynolds number varies from 10,000 to 10,000,000, K will only vary from 0.6 to 0.8 for an orifice plate and from 0.90 to 0.99 for a venturi.

For most liquids and gases using an orifice flowmeter or venturi, the flow equation is

$$W = 359 \times K \times d^2 \times \sqrt{\rho \times h_w}$$

where: W = flow, lb/hr
K = coefficient of discharge (see Table 2.1 for typical coefficients)
d = orifice diameter, inches
ρ = specific weight of fluid, lb/ft^3
h_w = differential pressure drop, inches of water

This equation is used in the design and application of an orifice flowmeter to find one of the three variables W, d, or h_w, given the other two variables. For example, if the diameter, d, of the orifice and the diameter, D, of the pipe are given, and the required maximum flow, W, is specified for the designated type of fluid, then the differential pressure range, 0 to h_w, corresponding to the flow range, 0 to W, can be calculated for the transmitter. The Reynolds number, R_D, can be calculated from its above equation. The beta ratio equal to d/D can also be calculated. With R_D and the beta ratio, the coefficient of discharge table (e.g., Table 2.1) can be interpolated to find K for the orifice. With ρ for the specified fluid, the flow equation can now be solved for h_w, the differential pressure to which the transmitter must be calibrated so that its signal has a range from 0 to W.

To use this equation to find an orifice design diameter, d, when the maximum flow rate, W, and the corresponding differential pressure, h_w, are specified, assume a value for K (0.65 for an orifice); find the approximate diameter, d, from the flow

*To review the tables covering the full range of pipe sizes, see American Society of Mechanical Engineers (1971).

TABLE 2.1
Partial table of coefficient of discharge, *K,* for flanged-tapped orifice flowmeters for 4-in. schedule 40 pipe (I.D. = 4.026″)

	$R_D \times 1000$								
ß	10	20	40	80	160	320	640	1280	2560
.1	.5965	.5963	.5963	.5962	.5962	.5962	.5962	.5962	.5962
.125	.5969	.5966	.5965	.5964	.5964	.5964	.5964	.5964	.5964
.150	.5974	.5970	.5968	.5967	.5967	.5967	.5966	.5966	.5966
.175	.5981	.5975	.5973	.5971	.5971	.5970	.5970	.5970	.5970
.2	.5991	.5982	.5979	.5977	.5976	.5975	.5975	.5974	.5974
.225	.6002	.5991	.5986	.5983	.5982	.5981	.5981	.5980	.5980
.250	.6016	.6002	.5995	.5992	.5990	.5989	.5988	.5988	.5988
.275	.6033	.6015	.6007	.6002	.6000	.5998	.5998	.5997	.5997
.3	.6053	.6031	.6020	.6015	.6012	.6010	.6009	.6009	.6008
.325	.6077	.6049	.6037	.6030	.6027	.6025	.6023	.6023	.6022
.350	.6105	.6072	.6057	.6049	.6045	.6042	.6040	.6040	.6039
.375	.6138	.6098	.6081	.6071	.6066	.6063	.6061	.6060	.6059
.4	.6176	.6130	.6108	.6097	.6091	.6087	.6085	.6084	.6083
.425	.6220	.6166	.6141	.6128	.6121	.6116	.6114	.6113	.6112
.450	.6271	.6207	.6179	.6164	.6155	.6150	.6148	.6146	.6145
.475	.6328	.6256	.6223	.6205	.6196	.6190	.6187	.6185	.6184
.5	.6394	.6311	.6273	.6254	.6242	.6236	.6232	.6230	.6229
.525	.6469	.6374	.6332	.6309	.6296	.6289	.6285	.6282	.6281
.550	.6554	.6447	.6398	.6373	.6358	.6450	.6345	.6342	.6341
.575	.6650	.6529	.6474	.6445	.6429	.6420	.6414	.6411	.6409
.6	.6759	.6623	.6561	.6529	.6511	.6500	.6494	.6490	.6488
.625	.6882	.6729	.6660	.6624	.6603	.6592	.6585	.6581	.6578
.650	.7022	.6851	.6773	.6733	.6710	.6697	.6689	.6684	.6681
.675	.7181	.6990	.6903	.6857	.6832	.6817	.6808	.6803	.6800
.7	.7363	.7148	.7052	.7001	.6972	.6955	.6946	.6940	.6937

equation; calculate R_D and $\beta = d/D;$ then interpolate the table for *K;* and then accurately recalculate *d* from the flow equation. Repeat for another iteration to find *d,* and if *d* differs from its previous value by less than 0.001 in., you have an accurate value for it—otherwise perform another iteration.

EXAMPLE 2.4 **Orifice Flowmeter Calculation**

Calculate the diameter, *d,* of a flange-tapped orifice flowmeter for a 4-in. schedule 40 pipe carrying water at 60°F. The flow transmitter range is calibrated for 0 to 100 in.

of water for the desired flow range of 0 to 150 GPM. Use Table 2.1, relating coefficient of discharge, K, to pipe Reynold's number, R_D, and ß ratio for 4-in. schedule 40 pipe with a flange-tapped orifice meter. The calculated result within 4 iterations is as follows:

Assume: $K = 0.65; \rho = 62.4 \text{ lb/ft}^3$; 1 USG = 8.33 lb of water
$\mu_1 = 2.1\times10^{-5} \text{ lb}_f - \text{sec/ft}^2$ (water @ 60°F)

First try:
$$d^2 = W / (359 \times K \times (\rho \times h_w)^{1/2})$$
$$= 150 \times 8.33 \times 60 / (359 \times 0.65 \times (62.4 \times 100)^{1/2})$$
$$= 4.06712$$
$$d = 2.017''$$

Second try:
$$d/D = \beta = 2.017/4.026 = 0.5010$$
$$R_D = W / (235.6 \times D \times \mu_1 \times g)$$
$$= 150 \times 8.33 \times 60 / (235.6 \times 4.026 \times 2.1\times10^{-5} \times 32.2)$$
$$= 116{,}886 = 117{,}000$$
$$K = 0.6250 \text{ from table for } \beta = .501, R_D = 117{,}000$$
$$d = 2.057''; \beta = .5108$$

Third try: $K = 0.6272; d = 2.053''; \beta = .510$

Fourth try: $K = 0.6270; d = 2.053''; \beta = .510$

Note that the diameter, d, will increase with an increase in temperature that causes it to expand. Also note that the fluid specific weight, ρ, varies with temperature, and for gases the specific weight also varies with pressure change. This latter variation means that, as the gas pressure drops while passing through the flow constriction, the gas density also drops. It is recommended that only the concentric, flange-tap, square-edged orifice flowmeters (these are the most common flowmeters) be applied.*

Vortex Shedding Flowmeter

The vortex shedding flowmeter became popular in the 1970s and is replacing the orifice flowmeter in many liquid, gas, and steam applications. It offers a much wider measuring range of accurate operation. For example, the orifice square root factor causes standard measurement errors in differential pressure to become drastic flow errors as the flow becomes smaller and smaller; an error of 1% in full scale differential pressure introduces an error of 0.5% in actual flow at 100% full scale flow rate, but it introduces an error of 2% of full scale flow at 25% of full scale flow rate. The vortex meter error is linearly related to flow rate. Linear ranges of flow with a turndown ratio of 30 to 1 are feasible with this instrument.

---

*For a detailed description with examples of the calculations needed to find the relationship of flow to differential pressure for orifice plates and venturis, see American Society of Mechanical Engineers (1971).

Vortices are shed from an obstruction in a flow stream. These vortices are rolling streamlines that alternately curve off the obstruction from one side to the other and roll downstream. They occur when the breeze blows past a flag and causes the flag to flutter. In a pipe the velocity of the vortices flowing in the stream is directly related to the stream velocity. The number of vortices passing a point in the pipe is related to this velocity. A device that counts the vortices passing per second will also measure the flow rate. Vortices are inhibited in viscous fluids at low flow rates. For this reason there is a Reynolds number (approximately 10,000) below which a vortex flowmeter will not be effective. At high fluid velocities the obstruction may introduce excessive pressure drop and this is usually the limit to higher flow rates. Care must be taken that liquids do not cavitate (vaporize under low absolute pressure) in the flowmeter, since this may cause damage to the sensitive vortex detector. The vortex shedding meter includes the obstruction, the vortex detector, and electronics used to transmit a flow signal, all packaged to mount in the pipe between flanges. Therefore the cost of installation requires no impulse tubing nor valve manifold, as required for an orifice flowmeter. Also, there is less need for concern about freezing, which can occur with impulse lines. Installation in a vertical or a horizontal pipe is acceptable for the vortex meter.

Standard digital and analog signals are available, especially 4–20 ma DC, with errors in the manufacturer's calibration less than 1% of actual flow rate.

Magnetic Flowmeter

The magnetic flowmeter (Figure 2.15) is ideal for liquids that conduct electricity. It is ineffective on gases and pure liquids such as oils that are not adequately conductive (less than 10 μsiemens/centimeter).

The conductive liquid forms an electrical conductor that obeys Faraday's Law of Electromagnetic Induction as it moves through the magnetic field established inside the flowmeter. This conductor in the magnetic field will generate an electric voltage that is proportional to its average velocity. The voltage is in the millivolt range and an amplifier is provided to convert the signal to 4–20 ma DC. Coils on the outside of the nonmagnetic, stainless steel pipe are pulsed (at about 5 Hz) D-C. The resulting pulsed magnetic flux flows initially through a laminated steel core to the top of the stainless steel pipe and then through the stainless steel pipe (and thus through the flowing liquid) to the other end of the laminated steel core and then back through the coils. Pulsed D-C allows these newer magnetic flowmeters to offer the advantage of compensation against electrical noise and extraneous magnetic fields on every pulse. (When the pulse is shut off, the background electromagnetic noise is measured and compensated for.) The older design, using line frequency A-C, excited magnetic fields that can only be manually compensated infrequently.

These instruments are usually designed for flange mounting with an insulating plastic pipe liner that extends out into the flanges. The electrodes extend in insulators through the pipe and are flush with the inside of the liner. These electrodes are perpendicular to the pipe and the magnetic field that is established by the coils wrapped

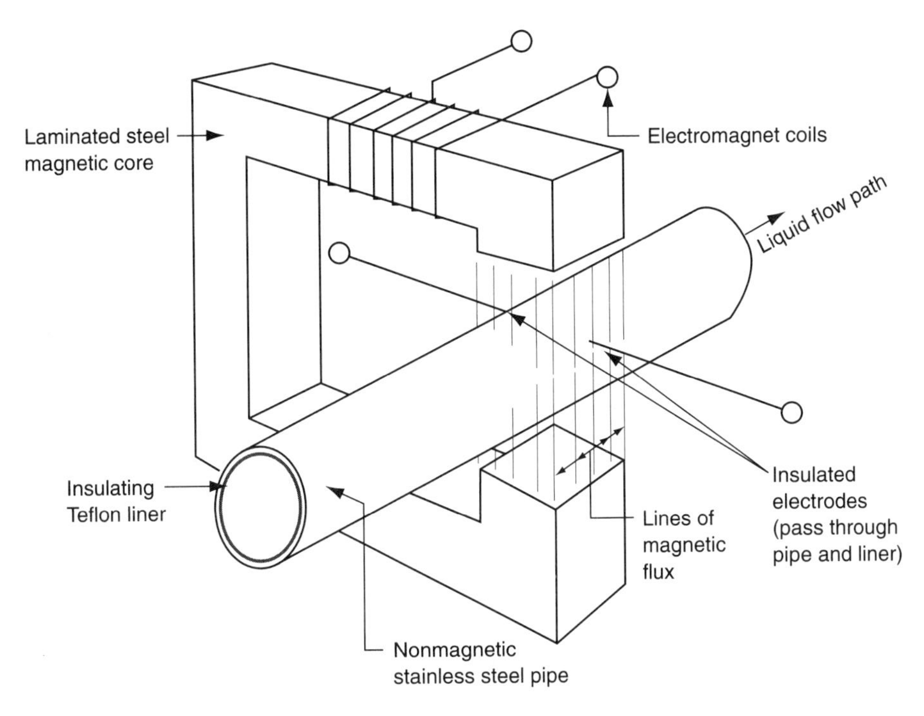

FIGURE 2.15
Magnetic flowmeter

around the pipe. A variety of methods are offered for cleaning coatings from the electrodes. Such coatings are due to process liquids that contain dissolved solids that precipitate on the electrodes. A ground strap is required to ensure that stray electrical currents in the pipe will pass from the pipe on one side of the meter to the pipe on the other side of the meter through a conductor other than the flowing liquid. It is also essential to ground the flowmeter body.

A magnetic flow measurement is not affected by viscosity, temperature changes, pressure changes, or flow direction, and it does not introduce any more pressure drop in the pipe than an equivalent piece of pipe. Its range of flow has a turndown of 100 to 1 with reported errors less than 1% of actual flow rate over a range of 10% to 100% full scale, and 0.1% of full scale from 10% down to 1% of full scale.

The magnetic flowmeter is particularly applicable to slurries, corrosive liquids, and sewage, and liquids requiring measurement in both directions.

Ultrasonic Flowmeter

Ultrasonic transducers are strapped on the outside of the pipe and transmit an ultrasonic beam across the pipe through the flowing liquid. One type measures the transit time for the sound to traverse the pipe at an angle of about 45° to the pipe axis, first with, and then against, the flow of liquid. Another type of ultrasonic flowmeter, using only one transducer for transmitting and receiving the beam, applies the Doppler principle of frequency change of the ultrasonic beam due to the liquid velocity.

Both types of ultrasonic flowmeter are affected by particles or bubbles in the liquid. The transit time flowmeter cannot tolerate many bubbles (less than 1%) because they attentuate the beam strength, but the Doppler flowmeter relies on bubbles or particles to reflect the beam back to the sensing device. Therefore, the bubbles or particles must be suspended in the liquid. Pipe materials affect the ability to conduct sound from the sending device into the liquid and back again.

A standard 4–20 ma DC signal is available with quoted errors less than 3% of span. More accurate results may be possible if an individual and accurate measurement of the pipe's inside diameter is used in the equation relating volumetric flow to velocity.

Turbine Flowmeter

The turbine flowmeter has blades that spin due to fluid passing over them. The speed of the rotor which holds the blades is linearly related to the flow rate, and the number of turns completed is related to the quantity of fluid that has passed through the meter. Usually a magnetic pickup coil, mounted on the outside of the nonmagnetic stainless steel spoolpiece containing the rotor and blades, detects each blade as it passes. Pulses from the coil are counted electronically and their rate is converted to a 4–20 ma DC signal for transmission or converted to digital communication.

Turbine flowmeters function well with liquids and gases, but they require maintenance at their bearings from time to time. Liquid turbine flowmeters can be calibrated to have errors less than 0.25% of actual flow rate over a range of 10% to 100% full scale flow rate, and gas turbine flowmeters can be calibrated to have errors less than 1% of actual flow rate over a range from 5% to 100% full scale.

2.6 LEVEL

The level of a liquid or of solid particles is usually measured with respect to some reference elevation. Usually a vapor space exists above the liquid level and air above the solids level. However, sometimes the level of an interface, where two liquids meet, or the level of solids in a liquid is required to be measured. Generally the level is required above the reference elevation, but sometimes it is required below the reference elevation. Often it is satisfactory to know if the level is greater than or less than a value sensed by a level switch.

The most common methods of sensing level are as follows and are described in this section*:

- Bubbler and differential pressure transmitter (see p. 41).
- Electrical capacitance: an insulated electrode extends from the top to near the bottom inside a tank. As the level of the liquid inside the tank rises, the combined dielectric of the liquid and the insulation around the electrode

*For how the different types of level sensors are classified, see Anderson (1980) and Liptak and Venczel (1982).

varies. Electronic circuits sense this variation as a change in capacitance and produce a signal proportional to the level.

- Electrical conductivity: an electrode uninsulated at its tip extends into a tank. When the liquid in the tank (it must be slightly conductive) rises to the tip of the electrode, an electronic circuit detects the change in conductivity and produces a signal.
- Differential pressure (see p. 40).
- Repeater (see p. 40).
- Displacer or buoyancy sensor (see p. 41).
- Radioactive: a source of radioactivity is arranged to radiate across a tank to a radiation detector. The higher up inside the tank that the liquid or solid exists, the more it blocks the radiation that reaches the detector. The change in radiation changes the signal from the detector.
- Paddle wheel: a vaned wheel is placed under the roof of a silo on the bottom of a vertical shaft. The top end of the shaft is continually being driven by a motor on the roof of the silo. If the solid material (e.g., grain) in the silo rises to surround the vanes of the wheel, then it will stop turning and a signal will indicate excessive level in the silo. The device is designed so that the shaft may be stopped for days without damage to itself.
- Tape and surface detector (see p. 42).
- Ultrasonic (see p. 42).
- Weight (see p. 42).
- Tilt: a tilt-type mercury switch is suspended from a cable over a pile of solids (e.g., stones from a conveyor). When the conveyor has carried enough stones, so that the pile of stones rises to where it displaces the tilt switch, it causes a circuit to detect the excessive level of the pile.

These devices are used to detect a level point by operating a switch or to sense a continuous range of level using a zero value that may be elevated or suppressed relative to the reference elevation. Certain devices can be used for liquids, and certain others can be used for solid particles. For any particular application, there is usually some problem associated with each device. The question becomes: which problem causes the least concern? These devices, their applications, and some of their associated problems are summarized in Table 2.2. Notice *ratholing* may occur in a solid powder in a silo, where a hole may run from the top of the silo to the bottom due to the powder sticking to the walls of the silo; *arching* may occur in the silo with a lot of the powder removed from under the arch.

Liquid Pressure Sensors

The pressure and differential pressure sensors described in section 2.4 are readily used for sensing the pressure head due to liquid contained in a tank. Although a tube or pipe can extend from the bottom of the outside of the tank to the sensor diaphragm,

TABLE 2.2
Level sensor application problems

Device	Liquids' Problems		Solids' Problems	
	Switch	Continuous Range	Switch	Continuous Range
Bubbler + D.P.	1, 6	1, 6	x	x
Capacitance	2, 3, 12	2, 3, 12	x	x
Conductivity	2, 3, 12	x	x	x
Diff Press (D.P.)	6, 7, 8	6, 7, 8	x	x
Repeater + D.P.	6	6	x	x
Displacer	9, 10	9, 10	x	x
Radioactive	4, 5	4, 5	4, 5	4, 5
Paddle wheel	x	x	11, 13	x
Tape + Surface Det	5, 10, 15	5, 10, 15	5, 13, 15	5, 13, 15
Ultrasonic	12, 14	12, 14	11, 13, 14	11, 13, 14
Weight	5, 9	5, 9	5, 9	5, 9
Tilt	x	x	9, wind	x

Problems associated with each device and application:

1. Solids buildup on bubbler tube
2. Signal nonlinearity
3. Liquid electrical resistivity
4. Hazardous to personnel
5. Expensive
6. Liquid density variations
7. Freezing of impulse lines
8. Wet leg not full
9. Installation
10. Turbulence of liquid
11. Ratholing, arching
12. Foam
13. Dust buildup
14. Temperature compensation
15. Corrosion

x Device not applicable

it is often desirable to flange mount the differential pressure sensor on the tank so that its diaphragm is flush with the inside of the tank and sediment will not plug the tube or pipe.

For tanks open to the atmosphere, the differential pressure is measured between the sensor connection point to the tank and the atmosphere. For pressurized tanks it is necessary to measure the differential pressure between the connection point to the tank and the vapor pressure above the surface.

If the center of the sensor diaphragm is not at the tank reference elevation, then the sensor has its zero suppressed or elevated. This way the zero-level transmitted signal (3 psig or 4 ma) will be at the pressure corresponding to the difference between the tank reference elevation and the sensor diaphragm center.

For tanks containing difficult to handle liquids, special pneumatic pressure repeaters can be flange mounted on the tank at the bottom as well as above in the vapor space. These devices sense the pressure inside the tank at the point of measurement and convert it to an equal pneumatic pressure, which can then be transmitted to the differential pressure sensor that measures the tank level.

A common technique used in measuring liquid level involves a bubbler tube. This tube is mounted in the tank so that its open end is downward near the bottom of the tank. Air is slowly bubbled down through the tube so that it contains no liquid and the air comes out the bottom of the tube in bubbles, about 1 or 2 bubbles per second. This way the air pressure in the tube is just enough to overcome the liquid pressure at the bottom of the tube; but the air pressure in the tube is the same from top to bottom of the tube, unlike the liquid pressure that changes significantly from top to bottom of the tube. A simple needle valve restricting plant air pressure at the entrance of the tube at the top of the tank is all that is needed. Some suppliers, however, provide a constant air flow device to maintain a constant flow of air regardless of the liquid level (and equivalent pressure) in the tank. The air pressure in the bubbler tube can be read directly with a gauge or pressure switch, or its value can be transmitted with a pressure transmitter or differential pressure (D.P.) transmitter to a remote location.

Sometimes two bubbler tubes are used on an enclosed, pressurized tank. One tube has its outlet point at the top in the vapor space and the other tube has its outlet point at the bottom, in the liquid. These bubbler tubes obtain the vapor pressure as well as the pressure at the bottom of the tank for a differential pressure transmitter. In this case it is necessary to ensure that the excess air does not interfere with the vapor.

For tanks containing boiling water, a special problem exists. The impulse lines connecting such a tank to the pressure measuring level sensor are normally at ambient temperature and thus contain liquid since the steam from the vapor space condenses in its higher impulse line; and in most cases the lower impulse line is filled with liquid. For a tank with steam, the high pressure side of the sensor is not connected to the liquid side but to the vapor side of the tank, and this impulse line is called a *wet leg*. Steam runs into it at the top, condensing, filling the wet leg, and liquid is overflowing back into the tank at the top. The level signal is reversed from normal. A low pressure means a high level and a high pressure means a low level. In this case special care must be taken if zero elevation or suppression is required.*

Liquid Buoyancy Sensors

As liquid rises around a float that is restrained from rising or falling, the buoyant force on the float increases. The float is usually referred to as a *displacer.* Measurement of the buoyant force gives a value of the liquid level. The area of the displacer must be constant along its vertical length, and its vertical length must be slightly longer than the change in level that must be measured. A displacer is a cylinder with its length selected to slightly exceed the level to be measured and its diameter selected to provide adequate buoyant force to operate the sensing mechanism.

Displacers are used to measure liquid level, interface level between two liquids, and liquid specific gravity (which requires the displacer to be fully submerged). Torque

*For a variety of examples of these kinds of applications, see Anderson (1980) and Liptak and Venczel (1982).

tube and force balance diaphragm seal mechanisms are used to transmit the buoyant force from inside the tank to the transmitting device (pneumatic or electronic) on the outside of the tank.*

Ultrasonic Sensors

Ultrasonic level sensors are proven reliable sensors for measuring the liquid level in a tank and the level of solid particles in a bin or silo. The continuous level sensor has a sound generator and receiver mounted above the liquid surface or above the pile of particles. A 20-kHz sound wave is beamed down to the surface, and the reflected signal returns to the receiver from the surface after a certain time. The further down the level is, the longer will be the time to receive the reflection. Ultrasonic devices may also be used for discrete liquid level measurement. Such devices have a tube with a notch in it that has a sonic generator on one side of the notch and a receiver on the other side of the notch. When the tank liquid rises above the notch, the sonic beam is transmitted through the liquid to the receiver; and when only vapor is in the notch, the sonic beam is weakly transmitted to the receiver so that no signal is detected.

The temperature of air significantly affects the velocity of sound waves. Ultrasonic level sensors must be compensated for air temperature changes.

Container Weight Sensors

To find the quantity of material in a tank, it is sometimes desirable to weigh the tank, although this is usually more expensive than measuring the level in the tank and using the level to infer the quantity. The most common method of weighing a tank is to place electric load cells in all the columns supporting the tank. These load cells detect the force in each supporting column, and the summation of all these forces is the weight of the tank including its contents. A major problem is the pipes connecting the tank to the rest of the process. They must be flexible so that they do not affect the load cells significantly.

Tape Plus Surface Sensors

Liquid level measurement, using a float that continually rides on the surface, is effective for clean liquids. One method uses a tape containing multiple electrical conductors arranged in a geometrical, binary, gray code pattern. The tape is attached from the bottom of the tank to the top of the tank. A transducer in the float, riding on the tape, induces a signal into the conductors near it at that level and digitally selects the conductors corresponding to the level. Fourteen conductors, plus a reference and a return conductor, resolve 135 ft into 0.1-in. increments.

Another method for liquid level uses a float maintained between wire guides running from the top to the bottom of the tank. A metal tape on the top of the float runs up to a pulley at the top of the tank and then through other pulleys to a head unit, where it is wound up on a reel. The tension in the tape is maintained constant by the takeup device so that the float rides on the surface of the liquid. When the

*Anderson (1980) and Liptak and Venczel (1982) describe how displacers are used.

tape moves up or down, a counter keeps track of its position. With special seals this unit can be used on pressurized tanks.

For solids level measurement, a sounder (similar to a float) on a tape is lowered intermittently (e.g., once every 20 minutes) into the silo until the sounder touches the surface of the solids and the tension in the tape is reduced. The tape is then rewound to a reference level, and the distance the tape rises is counted digitally. Dust buildup on the tape is a major problem with this device. The sounder should be well out of the way of any solids that are filling the silo.

2.7 FORCE AND WEIGHT

Weight is force. The force that the Earth's gravitational attraction applies to each body on its surface is called the *weight of the body.* The mass of each body is associated with the material of the body. If the distance of that body from the center of the Earth changes, then its attractive force toward the center of the Earth changes, and its weight changes even though its mass does not change. Also the weight of a body near a mountain may be slightly different from its weight on a plain because the gravitational force, due to the mass of the mountain, acts as a vector with the vector of the gravitational force due to the rest of the mass of the Earth. This is expressed as

$$\text{force (newtons)} = \text{mass (kg)} \times \text{acceleration (m/s/s)}$$
$$\text{weight (newtons)} = \text{mass (kg)} \times \text{acceleration due to gravity (m/s/s)}$$

The mass of a body remains constant. But if acceleration due to gravity changes, then weight changes. If force can be measured with a sensor, then obviously weight can be measured with a similar sensor. The acceleration due to gravity is specified at three locations on the Earth's surface as follows*:

Melbourne, Australia	9.79966 m/s/s
Foxboro, USA	9.80368 m/s/s
Soest, the Netherlands	9.81276 m/s/s

In order to display the same value of weight for a body of mass X at these three places, a transducer would have to be recalibrated slightly to compensate for the change in gravity.

Load cells are used for measuring all kinds of force, but weight is the most common kind of force. There are three kinds of load cells. Pneumatic and hydraulic load cells are designed almost exclusively for compression forces due to weight. Electrical load cells, however, are more universal and are designed to measure tension or compression forces.

Ambient conditions of temperature require compensation of the load cell signal. Ambient conditions of humidity, impact shock, vibration, or corrosion require protection of the load cells and associated platform weighing system. Provision for regular maintenance and calibration must be made. In some cases the law requires regular certification of the calibration.

---

*See page 38 of Anderson (1980).

Pneumatic and Hydraulic Load Cells

These cells have pistons and cylinders and are arranged so that, with no fluid (hydraulic or air), they rest on restraining metallic guides. The fluid lifts the cell piston only a few hundredths of a centimeter, thus the fluid pressure corresponds to the weight on the piston divided by the area of the piston. If several cells are used under a table to weigh the table and any body (such as a cement truck) resting on the table, then the fluid pressure signal from each cell is totalized with the signals from all the other cells to produce one signal.*

Electrical Load Cells

These devices use bonded strain gauges. The strain gauge is made of metal wire or metal foil or semiconductor material. The strain gauge is bonded to the unstressed element (i.e., metal strut or diaphragm to be placed in compression or tension) with epoxy cement. The strain in the element, when it becomes stressed, also strains (stretches or compresses) the strain gauge wire or foil. The change in the dimensions of the wire or foil changes its electrical resistance. The change in resistance is related to the change in strain that is related to the change in stress (pressure in the stressed element) and the force on the stressed element.

2.8 ANALYSIS (CHEMICAL) VARIABLES

Chemical analysis is an immense subject, and no one book could possibly cover it. The sensors used for chemical analysis are extremely complex. For example, a chromatographic analyzer, which analyzes a product for the amount of each chemical component, includes several microprocessors, specialized optical sensors, packed columns of specialized chemicals, and may cost over $100,000. Analyzers for specific gravity, viscosity, humidity, pH, and liquid electrical conductivity are fairly common, and they are relatively simple. Analyzers of combustion products have great value because they can significantly reduce fuel costs. And new families of analytical sensors are being developed. Some of them are very tiny and are based on biological functions. Biological activity is being studied to understand these very specific analyzers that animals and plants utilize.

In this section, humidity and electrolytic conductivity sensors are described.

Humidity Analysis

Humidity is the concentration of moisture in a gas, usually in the surrounding atmosphere, which is a mixture of gases. If we assume the absolute air pressure is constant at 14.7 psia, then the air will hold a maximum amount of moisture in a given volume of space, depending on the temperature of the air. The higher the temperature, the

*For more details, see Liptak and Venczel (1982) and Considine (1985).

more moisture the air holds. If the air is not saturated with moisture, then it will be holding somewhat less than the maximum that it could hold. So, at any particular pressure there is a saturation amount of moisture in a given volume of space, depending on the temperature of the air. If the air is holding the maximum or saturation amount in the given volume of space, then it has a *relative humidity* of 100%. If it is holding 50% of the maximum amount, then it has a *relative humidity* of 50%. Another humidity measurement is the *absolute humidity* of air, given by the mass of vapor held in a given mass of air, for example, 0.008 lb of moisture per 1 lb of dry air.

When a given volume of ordinary air holding some amount of moisture is cooled, then as some cool temperature is reached, a liquid dew forms on solid objects in the air. The temperature at which dew just starts to form is called the *dew point temperature*. At this temperature the air has reached and slightly exceeded 100% relative humidity and will not hold as much vapor, so it must condense the vapor into liquid onto solid objects.

The sling psychrometer is an instrument used to measure humidity. With the psychrometric chart (see Figure 2.16), this instrument can provide relative humidity (RH), absolute humidity, and dew point temperature. The sling psychrometer consists of two thermometers mounted together in a holder which has a handle that allows the operator to whirl the thermometers through the air. One thermometer has an ordinary dry bulb that measures the regular air temperature. The other thermometer has a wet bulb that is wrapped in a cloth wick soaked in water. When the psychrometer is whirled through the air, the dry bulb thermometer registers the regular air temperature. The wet bulb, however, evaporates water from the wick and in so doing removes heat from the bulb to lower its temperature. The amount that the temperature is lowered is dependent on the amount of vapor in the air. The smaller the amount of vapor, the more evaporation may occur and the lower the wet bulb temperature. After the sling psychrometer has been whirled for about one minute, the wet bulb temperature is quickly read and then the dry bulb temperature is read. With the dry bulb and wet bulb temperatures, the psychrometric chart can be used to find the relative humidity (RH), the absolute humidity, and the dew point (DP) temperature. On the psychrometric chart, the dry bulb temperature is found on the bottom horizontal scale and its vertical line is followed up to where it crosses the line sloping down toward the right that corresponds to the wet bulb temperature. At the point where they intersect, the relative humidity may be estimated from the scale on the curving lines. Starting at the same point of intersection, the dew point temperature is found where the horizontal line meets the 100% RH curving line. It is read from the dry or wet bulb temperature; they both have the same value at 100% RH. The absolute humidity in pounds of moisture per one pound of dry air is found by continuing horizontally from the dew point to the left side vertical scale.

Several types of sensors exist to measure humidity. The absolute humidity and the dew point temperature are directly related to one another as can be seen on the psychrometric chart since the horizontal line establishes the dew point and the absolute humidity. Any sensor that measures only dew point or absolute humidity also needs a measurement of air temperature in order to obtain relative humidity. The common types of humidity sensors include:

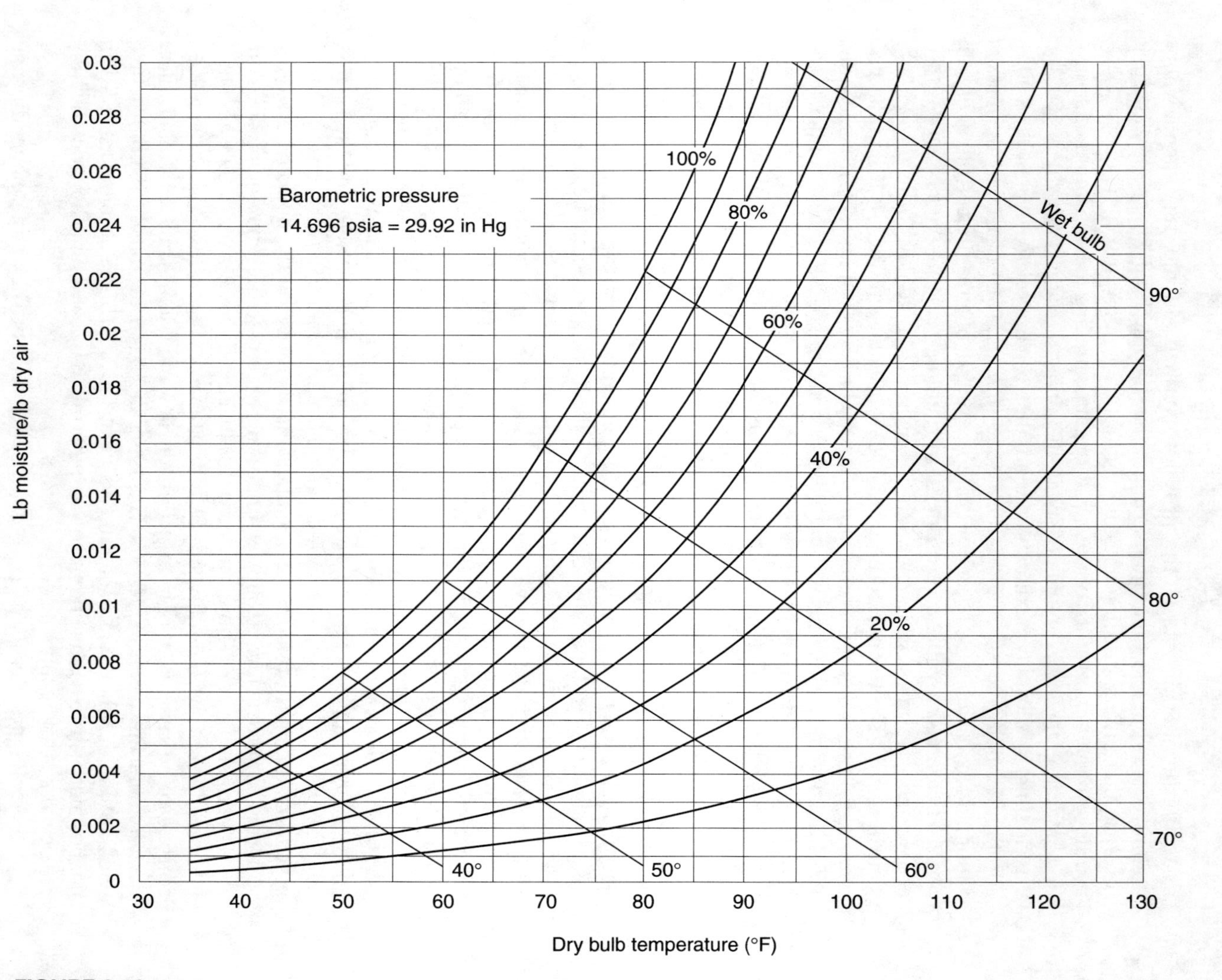

FIGURE 2.16
Psychrometric chart

- Wet and dry bulb psychrometer
- Lithium chloride element
- Chilled mirror

The lithium chloride element is a hollow, insulated spindle coated with a lithium chloride salt. Gold wires are wrapped over the salt so that electricity can pass through the wires into the salt and heat the salt and the spindle. As the salt heats up, it drives off moisture and the electrical resistance of the salt increases. As a result, the current through the salt decreases and the spindle cools down a little and absorbs more moisture at the cooler temperature. The temperature will oscillate a little at first, but after about ten minutes it will settle to an equilibrium value. A thermometer or temperature sensor placed inside the hollow spindle is used to record this internal equilibrium temperature. The internal temperature is directly related to the dew point temperature as shown in Figure 2.17. This sensor may be placed in a duct

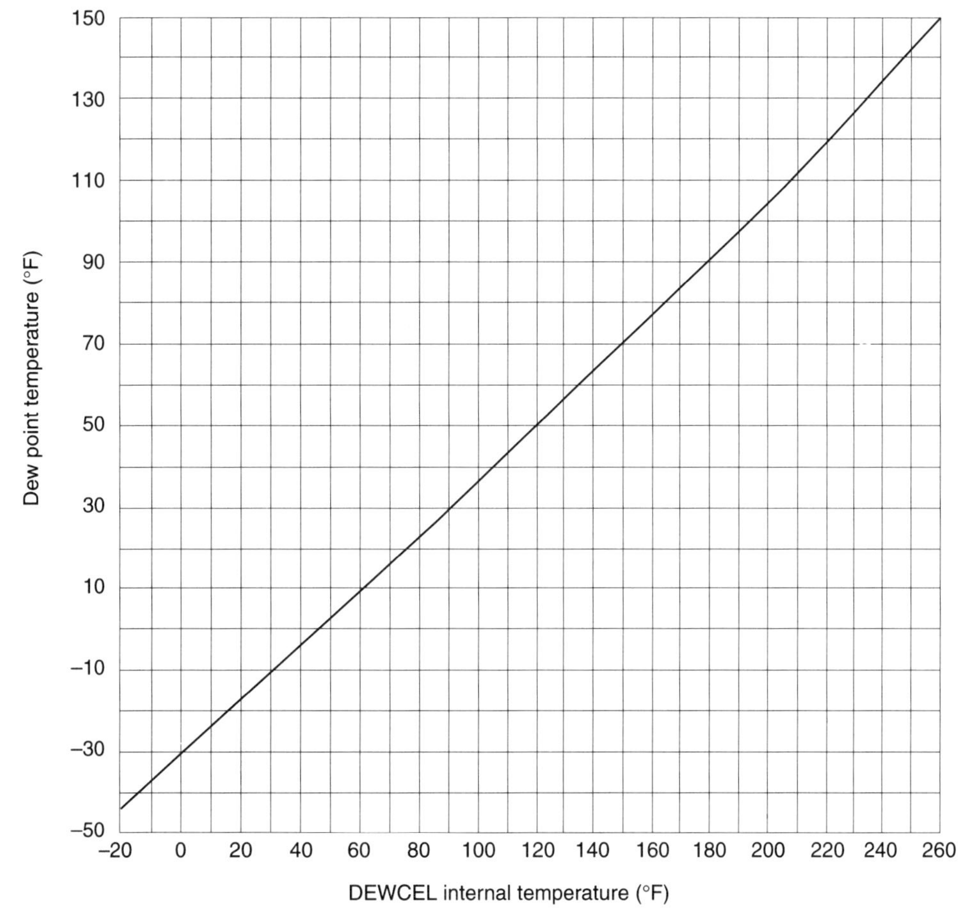

FIGURE 2.17
DEWCEL operating characteristics *(Courtesy: Foxboro Canada)*

to measure the humidity in flowing gas, such as air. It does not work well with gases that react with lithium chloride salt crystals or water; these gases include ammonia, sulphur dioxide, and chlorine. Using Figures 2.16 and 2.17, the relative humidity can be found from the DEWCEL internal temperature plus the (dry bulb) air temperature. The dew point temperature is found from Figure 2.17 using the DEWCEL internal temperature. The end of the horizontal line of the psychrometric chart (Figure 2.16) starting from the dew point temperature on the 100% RH line and ending at the dry bulb air temperature vertical line establishes the relative humidity, which can be estimated from the scale of the curving lines.

The chilled mirror sensor contains a mirror whose temperature is cooled by a thermoelectric cooler. A photocell detects the reflection of light from the mirror and adjusts the cooler to allow a small amount of light to be reflected when dew starts to appear on the mirror. The temperature of the mirror is measured with a temperature sensor, and this is the dew point temperature.

Electrolytic Conductivity Analysis

Electrolytic conductivity specifies how well electricity is conducted through a liquid, and it depends on the chemical makeup of the liquid and its temperature. Electrical resistance of a wire is equal to the wire's length, l, in cm, divided by its cross-sectional area, A, in cm^2, all multiplied by the material's resistivity constant, ρ, in ohm–cm^2/cm. The formula is:

$$R \text{ (ohms)} = \frac{\rho \text{ (ohm} - \text{cm}^2\text{/cm)} \times l \text{ (cm)}}{A \text{ (cm}^2\text{)}}$$

The longer the wire, then the greater the resistance; the larger the cross-sectional area, then the smaller the resistance. In the same way the conductance, G, which is the reciprocal of resistance, of the wire is given by the formula:

$$G \text{ (siemens)} = \frac{1}{R} = \frac{\gamma \text{ (siemen} - \text{cm/cm}^2\text{)} \times A \text{ (cm}^2\text{)}}{l \text{ (cm)}}$$

Here, γ is the material's conductivity constant, and it is the reciprocal of ρ.

Electrolytic conductivity of a liquid is found in the same way as electrical conductivity of a wire. For example, if a glass container has an inside base of 1 cm by 1 cm, and the base has an electrode (also 1 cm by 1 cm), and liquid 1 cm deep is placed above the electrode, and another electrode 1 cm by 1 cm is just touching the top of the liquid, then we have an electrolytic conductivity cell. The electrodes are connected by copper wires to a conductance measuring instrument (an ohmmeter). Since A is 1.0 cm^2, and l is 1.0 cm in length, then the conductivity, γ, of the liquid in this cell will be equal to the conductance, which is the reciprocal of the resistance.

Usually the conductivity constant has a very low value, and the measurement is given in μsiemen – cm/cm^2 or μsiemen/cm. (Before 1960 this was called μmho/cm, because a mho is also the reciprocal of the ohm.) Since conductivity (and resistivity) vary significantly with temperature, the temperature must be measured to obtain a precise value of conductivity.

The container with electrodes for measuring electrolytic conductivity is called a *conductivity cell.* The dimensions of the cell are designed to give the cell a value called the *cell constant.* The value of the cell constant is equal to the the distance, *l,* in centimeters, between the electrodes, divided by the area, *A,* in cm^2, of one of the electrodes (each electrode is assumed to have the same area, and the back of the electrode is assumed to be insulated from the liquid). The value of the cell constant equals l/A.

$$\gamma\,(\text{siemen} - \text{cm/cm}^2) = \frac{G\,(\text{siemens}) \times l\,(\text{cm})}{A\,(\text{cm}^2)} = \text{conductance} \times \text{cell constant}$$

The conductivity is often used to provide information about the concentration of dissolved solids in the electrolyte. For example, boiler operators must use quite pure makeup water to produce steam from their boilers; otherwise they will soon have excessive solids in their boilers and these solids may coat the tubes in the boilers and make them less efficient for heat transfer. Therefore they continually monitor the conductivity of the water entering the boiler and the water that they drain from the boiler. For this they use conductivity cells that measure the conductivity of the liquid passing through them.

Human Sensors

Humans can gather data from many sources and sensors. They can read dials on old-fashioned measuring devices, they can perform chemical tests, they can receive reports via radio, and they can enter this information via a keyboard into a computer system.

2.9 ELECTRIC POWER INDUSTRY VARIABLES

Electric variables are sensed when monitoring an electric power system or when monitoring the performance of electrically operated process machinery or as secondary variables to infer a primary process variable.

For electric power systems and electrical machinery, the system frequency is maintained almost constant (at 60 Hz in North America and 50 Hz in most of the rest of the world). Instrument transformers are used for safety and for the economy of smaller wires and simpler insulation. A potential-type instrument transformer (abbreviated P.T.) is used when sensing a voltage greater than 120 V and a current transformer (C.T.) is used when sensing a current greater than 5 A. Therefore, in addition to the basic range of the sensor, there is a scaling factor for any potential transformer or current transformer.

EXAMPLE 2.5 **Watt Transmitter Application**

The watt transmitter shown in Figure 2.18 has an output signal of 4–20 ma DC corresponding to –100 kW (power flowing from the fan drive motor to the supply) to 0 to +100 kW (power flowing to the motor from the supply).

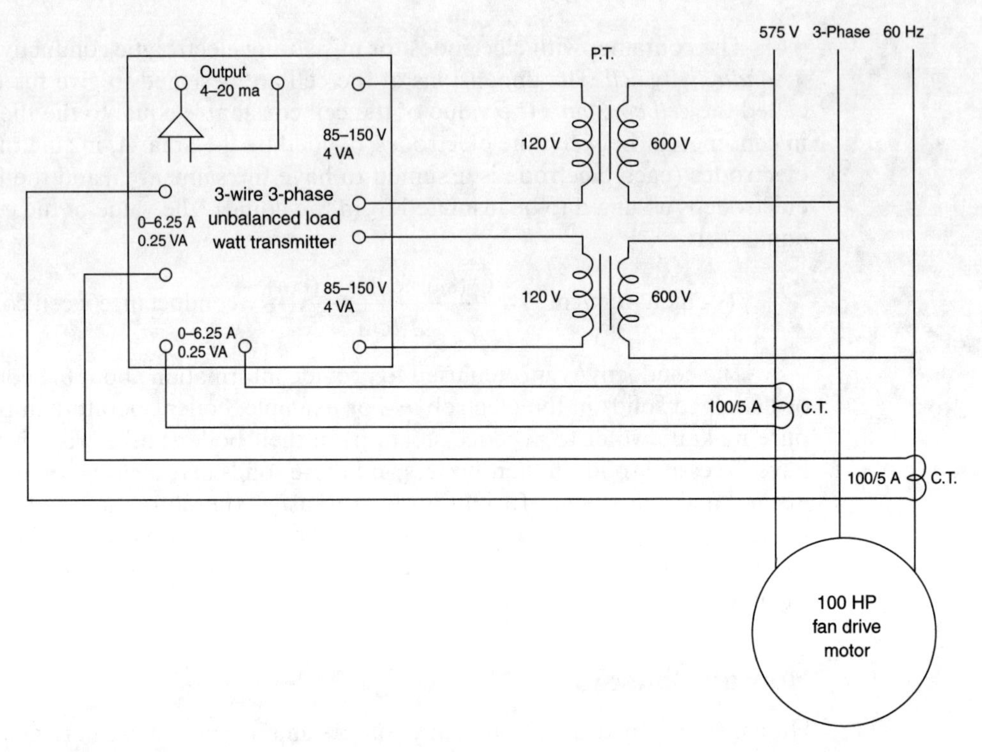

FIGURE 2.18
Watt transmitter

The transmitter is factory calibrated for 1000 W at a nominal 3-phase line voltage of 120 V and line current of 5 A to give 4 ma at –1000 W flowing from the motor to the supply, 12 ma at 0 W flowing, and 20 ma at +1000 W flowing to the motor from the supply. The potential transformers and current transformers have scaled the actual load power down to these values for each phase by (600 V/120 V) × (100 A/5 A) = 100 times. Therefore if each phase has 76.3 A at 575 V line-to-line at a power factor of 0.829, then the real power flowing to the motor = $\sqrt{3} \times 76.3 \times 575 \times 0.829 = 62{,}993$ W; and the transmitter will receive $5 \times 76.3 / 100 = 3.815$ A at $120 \times 575 / 600 = 115$ V line-to-line per phase. This means the transmitter receives $\sqrt{3} \times 3.815 \times 115 \times 0.829 = 629.9$ W and transmits $12 + 629.9 \times (20 - 12) / 1000 = 17.04$ ma when the motor is receiving 62.9 kW at a power factor of 0.829.

Voltage and Current

Self-powered, single-phase voltage and current transmitters provide 0–1 ma DC (or optionally 0–5 ma DC) corresponding to 0–150 V or 0–5 A at line frequency (within 50–500 Hz) with errors rated less than 2% of full scale. These devices respond to the average value of the amplitude of the AC signal, but they are calibrated for the rms or effective value of a pure sine wave, so distorted sine waves produce significant errors.

Vars (Volt-Amperes Reactive)

Var transmitters similar to the watt transmitters are also available. They are designed to calculate EI × sin (power factor angle) rather than EI × cos (power factor angle), which the watt transmitters calculate.

Phase Angle

Self-powered or externally powered phase angle transmitters provide –1 to 0 to +1 ma DC corresponding to 60° leading to 0° to 60° lagging (or optionally 90° leading to 0° to 90° lagging or 180° leading to 0° to 180° lagging) phase angles between a nominal voltage of 120 (85–135 V) and a current of 0.25–5 A. Errors are rated less than 2% of full scale.

Frequency

Self-powered frequency transmitters provide 0–1 ma DC or –0.5 to 0 to +0.5 ma DC corresponding to 45–55 Hz or 55–65 Hz or 380–420 Hz for 85–135 V signals. Errors are rated less than 2% of nominal frequency.

2.10 OTHER SENSORS

There are many other types of sensors that have not yet been mentioned in this book. Some of the most important are associated with measuring position, velocity, and acceleration. Other important sensors are associated with the analysis of physical conditions and chemical components of matter.

Position, Velocity, and Acceleration

Discrete manufacturing is the term given to production of machined parts. This type of processing makes use of numerical control and flexible manufacturing. The machinery used for such processing relies on sensors for position, displacement, motion, and object detection. The speed of motors is an important measurement using various types of tachometers. Machine vision is a new field that will involve an array of new technologies for object detection.

Aircraft and submarine navigation equipment use specialized sensors that integrate the craft's acceleration and velocity to obtain its change in position on the globe. Surveyors use laser techniques to obtain very accurately the position of buildings, roads, railway lines, mountains, and the contours of the hills that make up our world.

Vibrations are indicators of the health or the deterioration of health of rotating equipment. A family of sensors and a technology of vibration has been developed for sensing inertial acceleration, analyzing it, and presenting the information in a concise way.

2.11 DYNAMIC CALIBRATION

Steady-state calibration is dynamic calibration at extremely low frequencies. The result of a dynamic calibration is a Bode plot. A typical Bode plot is shown in Figure 2.19. On a logarithmic scale, the upper graph shows gain in output units per input unit as the sine wave frequency of the input varies from a very low value (almost DC or nearly 0 radians/second) to a high value, where the gain drops off significantly. For example, for the pneumatic-to-current converter the gain is in ma/psi, and at low frequencies (less than 0.01 Hz) it is (20 – 4) / (15 – 3) = 1.3333 ma/psi. The lower line of Figure 2.19 shows the linear phase shift in trigonometric degrees on a linear scale as the sine wave frequency varies over the same band as before. At low frequencies there is almost no phase shift, and at higher frequencies it reaches or exceeds 180 degrees.

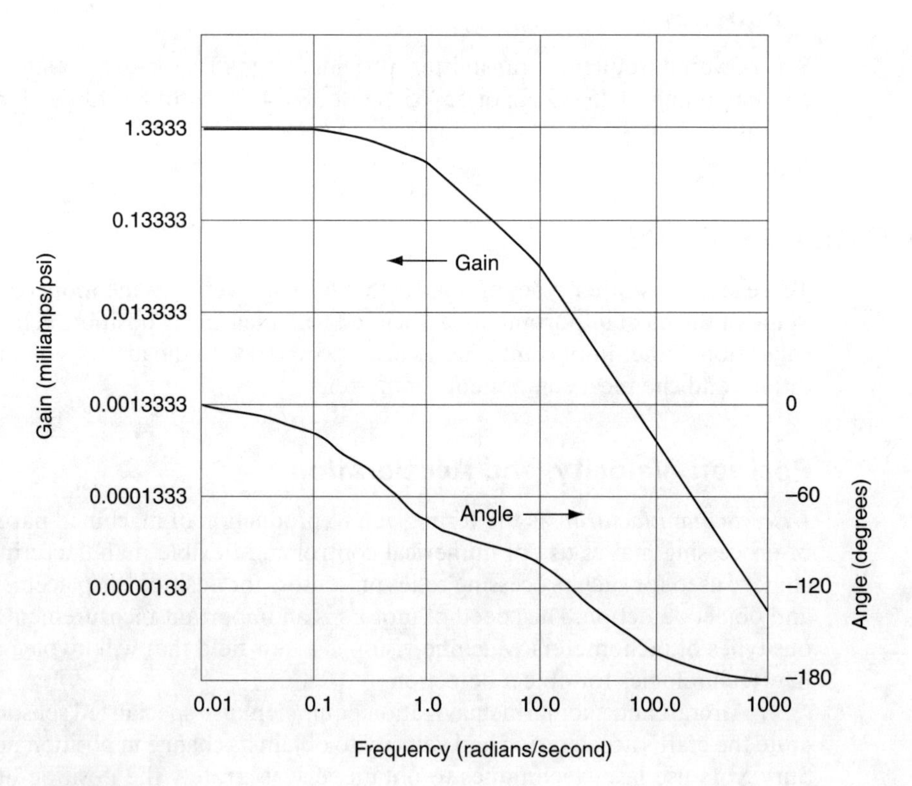

FIGURE 2.19
Typical Bode plot

A dynamic calibration is not apt to change significantly as time goes by, so it need only be done once. It is not apt to be significantly different from one device to another similar device. Therefore it is usually necessary to perform a dynamic calibration only once for a specific model of a device. The procedure for obtaining a Bode plot of a device is to excite the input signal to the device with a sine wave. The average value (or DC component) of the sine wave should be at about the middle of the input range and the alternating amplitude of the sine wave should not go outside the 20% to 80% points on the input range. The average then would be between 45% and 55% of the range, and the alternating amplitude would be about 10% to 25% of range. The frequency band should go from very low values to values where the phase shift is at least 85 degrees; if it will go beyond 90 degrees, then take the frequency up to at least 170 degrees of phase shift. You can display the input and output sine waves on a two-channel oscilloscope as shown in Figure 2.20 and record the peak-to-peak values in volts (where gain is the ratio of the output peak-to-peak value, *Ao,* divided by the input peak-to-peak value, *Ai*) for each channel as the frequency is scanned. You can also record phase shift in trigonometric degrees by measuring the distance (centimeters) that the peak of the output lags behind the input peak and the distance corresponding to 180 degrees peak-to-trough of either the input or output sine wave as the frequency is scanned. *Note:* 4 cm = 180° = π radians in a time of 4 × the sweep setting (sec/cm), thus the frequency (radians/sec) is π / (4 × sweep setting).

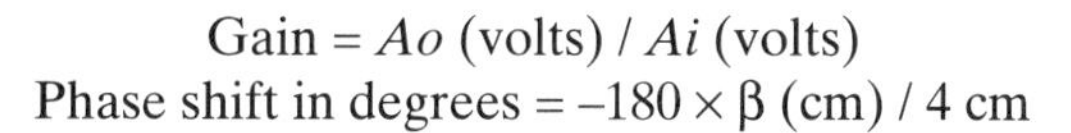

$$\text{Gain} = Ao \text{ (volts)} / Ai \text{ (volts)}$$
$$\text{Phase shift in degrees} = -180 \times \beta \text{ (cm)} / 4 \text{ cm}$$

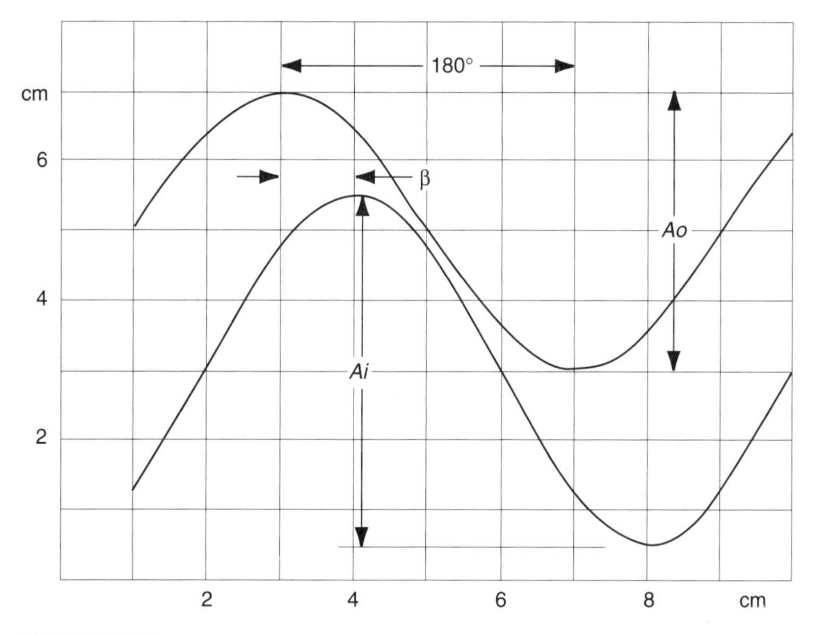

FIGURE 2.20
Oscilloscope display

EXAMPLE 2.6 **Frequency Response: Process Measurement**

Objectives: To measure and plot the gain and the phase shift of a demonstration process over the frequency band where they change significantly.

Equipment: Flow measurement demonstration process, low-frequency sine wave generator, storage tube oscilloscope (for very low frequencies), I/P converter and indicators.

Procedure: The demonstration water flow process is set up as shown in Figure 2.21.

With the process liquid flowing at about 50% of FT-1 flow range, set the sine wave amplitude to about 10% to 25% of FT-1 flow range. Start with a low frequency, where the gain between Channel A and Channel B is high and the phase shift is less than 5 degrees. Collect data for the columns in Table 2.3 of: Frequency in Hertz;

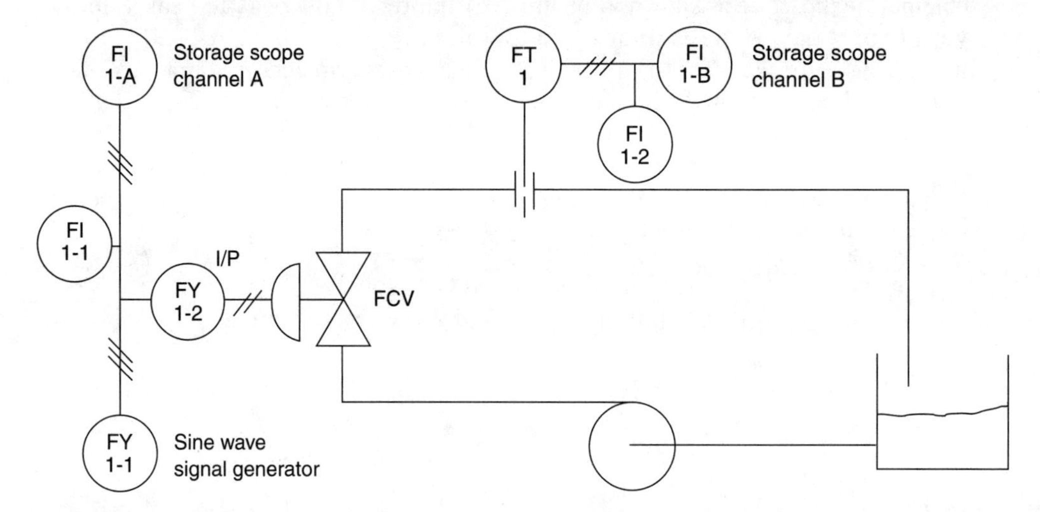

FIGURE 2.21
Water flow process

TABLE 2.3
Demonstration data for Example 2.6

Frequency		Peak-to-Peak					Wave Width		
Radians/s	Hz FY 1-1	Input (V) FI 1-A	Input (ma) FI 1-1	Output (V) FI 1-2	Output (ma) FI 1-2	Gain (ma/ma)	180° (cm)	Phase Shift (cm)	Phase Shift (in degrees)
0.06	0.01	2.1	11.7	4.06	14.2	1.21	10	0.1	1.8
0.126	0.02	2.1	11.7	4	14	1.19	5	0.5	18
0.251	0.04	2.1	11.7	3.86	13.5	1.15	6	1.5	45
0.503	0.08	2.1	11.7	2.43	8.5	0.72	6	2.8	84
1	0.16	2.1	11.7	1.43	5	0.43	3	2	120
2.01	0.32	2.1	11.7	0.43	1.5	0.13	3	3.4	204

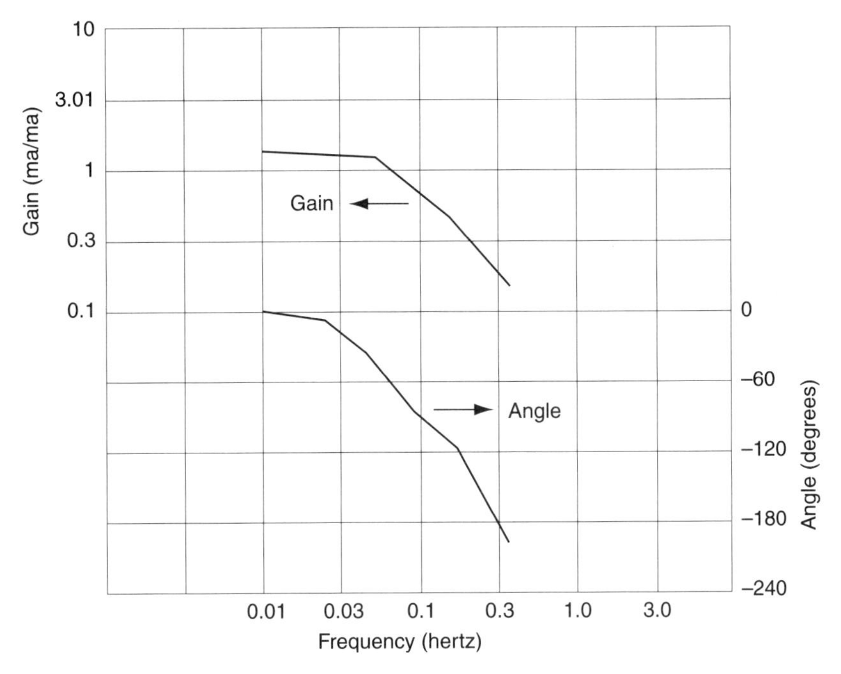

FIGURE 2.22
Frequency response

Peak-to-peak milliamperes and volts for the input (Channel A) and output (Channel B); Wave width in cm for 180° and Phase shift in cm. Double the frequency and repeat the data collection. Double it again and repeat. Continue doubling the frequency and collecting data until the phase shift is greater than 200° or the phase shift stays almost constant at 90° or 180°. Usually a dozen readings are enough. Calculate gain in output ma/input ma. Calculate phase shift in trigonometric degrees by taking the ratio of the phase shift in centimeters to the centimeters for 180° and multiplying by 180. Calculate the frequency in radians/second.

Conclusions: Plot the frequency response for gain and phase shift as shown in Figure 2.22.

PROBLEMS AND LAB ASSIGNMENTS

2.1 A technician calibrating a 0–50 psig pressure transmitter finds that the instrument has the following actual output milliampere signals: 4.233, 7.907, 12.197, 15.572, and 20.111, which correspond to the following standard psig input values: 1.008, 13.108, 26.520, 37.175, and 50.000. Prepare an "as-found" instrument calibration certificate and draw the corresponding instrument calibration graph and deviation graph.

2.2 The technician adjusts the transmitter of problem 2.1 to improve its calibration and the "as-calibrated" actual output milliampere signals become: 4.008, 8.011,

11.979, 16.211, and 19.911, which correspond to the following standard psig input values: 0.000, 12.518, 25.011, 38.002, and 50.000. Prepare the "as-calibrated" instrument calibration certificate and the calibration and deviation graphs.

2.3 Fill in all the columns of a calibration certificate for "as-found" data for a 0–100 psig pressure transmitter. The actual output data were 4.1, 7.9, 12.2, 15.8, and 20.1 ma, which correspond to standard input data of 2.0, 26.0, 52.0, 74.0, and 100.0 psig.

2.4 Use thermocouple tables in Appendix F for type J, iron/constantan, to find the temperature that exists at a type J thermocouple hot junction if the reference junction is at 0°C and the data acquisition system is reading 16.93 mV.

2.5 Use thermocouple tables in Appendix F for type K, chromel/alumel, to find the millivolts that are provided to a data acquisition system by a type K thermocouple at 229°C if the reference junction is at the ambient temperature of 20°C. If the ambient temperature increases by 5°C and the millivolts read by the system becomes 10.853 mV, what is the hot junction temperature?

2.6 Use thermocouple tables in Appendix F for type J, iron/constantan, to find the millivolts that will be provided to a data acquisition system by a type J thermocouple at 180°C if the reference junction is at the ambient temperature of 15°C.

2.7 Use thermocouple tables in Appendix F for type K, chromel/alumel, to find the millivolts that will be provided to a data acquisition system by a type K thermocouple at 180°C if the reference junction is at the ambient temperature of 15°C.

2.8 A constant value of 1.500 ma is applied to a Platinum RTD with a DIN curve of 43760, 9–68 in Appendix F. A data acquisition system measures the voltage drop across the RTD and finds 327.9 mV. What is the temperature of the RTD?

2.9 Describe a temperature sensor that you believe would be satisfactory to provide a 4–20 ma signal corresponding to a temperature range of 0 to 200°C. The process fluid is pressurized water flowing in a 3-in. pipe. Describe mounting methods and any transmitter required for use with the sensor. Obtain the specification from a vendor's catalogue.

2.10 Select the temperature sensor that you believe is best suited to measure the temperature of molten cement inside a rotating cement kiln. The maximum temperature is expected to be 1500°C. A signal of 4–20 ma is required. What range would you select? How would you mount this sensor. Obtain the specification from a vendor's catalogue.

2.11 A pressure transmitter has a range of –30 kPa to 70 kPa gauge pressure. What is its equivalent range in psia, in inches of water gauge, and in psig?

2.12 An automotive engine research establishment wishes to monitor the pressures inside a cylinder of a gasoline engine while the engine is running. They want to see the pressure changes inside the cylinder during each explosion. Select the best pressure sensor for this application and give your reasons for the selection.

2.13 Select the pressure sensor that you believe is best suited to measure water pressure at the base of a surge tank. The base is located 80 ft above the ground. The tank extends 40 ft above its base. The preferred location to view the surge tank pressure reading is 4 ft above the ground. Specify the range and type of sensor that you select. An output signal of 4–20 ma is required. Obtain the specification from a vendor's catalogue.

2.14 A flow transmitter measures gasoline into a customer's tank. From a time of 14:10:06 to 14:12:08, the flow rate is 0 liters/minute (lpm); from 14:12:08 to 14:12:21, the flow rate increases as a linear ramp to 173 lpm. It remains constant at 173 lpm until 14:16:48, when it is decreased at a linear ramp to 41 lpm at 14:16:55. It is then slowly shut off as a linear ramp until 0 lpm is reached at 14:17:23. How much gasoline was put into the customer's tank?

2.15 Determine the differential pressure range required for a differential pressure transmitter used to measure flow rate with an orifice plate. The process fluid is water at 60°F. The required flow range is 0–50,000 lb/hr. The orifice diameter is 1.75 in. The pipe size is 4-in. schedule 40. Use data from Example 2.4.

2.16 Describe the basic operation of a vortex flowmeter. What is the main advantage of the vortex flowmeter over the orifice flowmeter?

2.17 Describe the basic operation of a magnetic flowmeter. What are the two main advantages of the magnetic flowmeter over the orifice flowmeter? What is the reason for pulsing the magnetic flux to a magnetic flowmeter with direct current pulses at five pulses per second?

2.18 Select the most accurate type of flowmeter with the least, anticipated, maintenance problems to measure the flow of an abrasive liquid slurry in a horizontal pipe. Give your reasons for this selection. Does this flowmeter work well if the line is not full, but has air above the liquid slurry? If it does not work well, how can you get around the problem?

2.19 Draw a diagram of an orifice flowmeter measuring steam flow in a horizontal line. Show the orifice plate in orifice flanges and show the electronic differential pressure transmitter correctly connected to the orifice flowmeter with a three-valve manifold.

2.20 Select the most appropriate type of flowmeter with the least anticipated maintenance problems to measure the flow of a pure oil flowing in a 6-in. stainless steel pipe. Give your reasons for making the selection from the following flowmeters, explaining why it is better than the others:

Ultrasonic transit time

Ultrasonic doppler

Orifice plate

Vortex

Magnetic

Turbine

2.21 A pressure transmitter connected 10 cm above the bottom of a tank sends a signal of 14.38 ma to a computer. The transmitter was calibrated for a range of 0–100 kPa to produce 4–20 ma. If the liquid has a specific gravity of 1.0375, what is the level of the liquid in the tank? If the maximum expected error in the transmitter is 0.25% of span, what is the maximum expected error in centimeters?

2.22 If the tank in problem 2.21 is circular with vertical sides and a diameter of 10 m, what is the volume of the liquid and how much does it weigh?

2.23 If the tank in problem 2.22 is drained with a flow control loop set on automatic control at 822 kg per minute, how long will it take from the time the drain is started until the level reaches 3.33 m above the bottom? During the drain period no flow into the tank will occur.

2.24 A differential pressure transmitter with 4–20 ma output is used to measure the level of water in a tank located on a platform above the roof of a factory. The differential pressure transmitter is mounted under the tank below the roof. Its measuring point is 50 in. below the zero point of the tank. The tank is 95 in. high. Draw a diagram showing the tank, the roof, and the differential pressure transmitter, and the dimensions already described. What are appropriate values in units of pressure for the zero, span, and range? Is the zero elevated or suppressed?

2.25 Select the most appropriate level measuring instruments for each of the applications listed below. State your reasons for making each selection. The applications are as follows:

a. A continuous signal is required over the range 0–30 m for flour as it builds up in a silo;
b. A trip signal is required to shut off the conveyor bringing cement to a silo if the level of cement is within 1 m of the ceiling in the silo;
c. A continuous signal is required over the range of 0–10 m in a tank which is open to the atmosphere. The tank contains a foamy liquid which is very turbulent during the frequent filling periods.

2.26 Select the most appropriate type of continuous-range level sensor for a cylindrical tank lying on its side and containing a pure, foamy, turbulent petrochemical liquid with a specific gravity of 0.85. Give your reasons for making the selection from the following level meters, explaining why it is better than the others:

Ultrasonic

Differential pressure transmitter

Displacer

Radioactive

Paddle wheel

2.27 A sling psychrometer is used to measure humidity and shows a dry bulb temperature of 72.3°F and a wet bulb temperature of 55.5°F. What are the values of relative humidity in percent, absolute humidity in pounds of water per pound of dry air, and dew point in °F?

2.28 A sling psychrometer is used to measure humidity and shows a dry bulb temperature of 77°F and a wet bulb temperature of 60°F. What are the values of relative humidity in percent, absolute humidity in pounds of water per pound of dry air, and dew point in °F?

2.29 What is the temperature inside a Foxboro DEWCEL that is measuring the same humid air as the sling psychrometer of problem 2.27?

2.30 In certain humid air, a Foxboro DEWCEL registers an internal ambient temperature of 120°F. The temperature of the air is 65°F. What is the relative humidity of the air? What is the absolute humidity of the air?

2.31 A conductivity cell is used to measure the concentration of a salt (NaCl) solution at 18°C. The cell is made up of two square platinum electrodes. The left hand electrode has a clean, exposed external surface of 2 cm by 2 cm. Its back surface and edges are insulated with plastic. The right hand electrode has a clean, exposed inside surface, also 2 cm by 2 cm, and its back surface and edges are insulated with plastic. These exposed surfaces are maintained 2 cm apart. An electrical indicator used with this cell shows a conductance of 300 μsiemens. What is the conductivity of this electrolyte?

2.32 A conductivity cell is used to measure the concentration of a salt (NaCl) solution at 18°C. The cell is made up of two concentric platinum electrode rings. The inner ring has a clean, exposed external circumference with a diameter of 1 cm. The inner circumference and edges of this ring are insulated with plastic. The outer ring has a clean exposed internal diameter of 1.5 cm. The outer circumference and edges of this ring are insulated with plastic. The height of each ring is 0.25 cm. The indicator used with this cell shows a conductance of 118 μsiemens. What is the conductivity of this electrolyte?

2.33 Draw an electric supply circuit for a 60 Hz, 2500 HP, 3-phase air fan drive motor operating at 97% efficiency and 0.9 power factor on 4300 V showing two current transformers, two potential transformers, and a 3-phase, 3-wire unbalanced load watt transmitter rated –1000 to 0 to +1000 W at 5 A and 120 V. Show the transformer ratios that you select and give the range in kilowatts of the 4–20 ma output from the transmitter.

2.34 A large electrical load is connected to a 575 V single-phase supply. A 600/120 V potential transformer and a 100/5 A current transformer are placed in the supply lines to the load. A wattmeter is connected to these two transformers and it shows a reading of 673 W. Draw a diagram of the circuit showing the supply, transformers, load, and wattmeter. What is the actual power flowing to load?

2.35 What is the gain and phase shift of the pneumatic-to-current converter shown on the Bode plot of Figure 2.19 at 0.15 radians/second and at 0.25 Hz?

2.36 What is the frequency in radians/second of the output sine wave in the oscilloscope display of Figure 2.23? The grid divisions of the display are in centimeters, and the sweep range was set to 2 seconds/division. Channel 2 was set to a vertical deflection of 0.5 V/division and was driven from a 0–100 GPM flow sensor providing 4–20 ma into a 250 Ω 0.1% tolerance resistor. Channel 1 was set to a vertical deflection of 1 V/division and was driven by 4–20 ma passing through a 500 Ω 1%

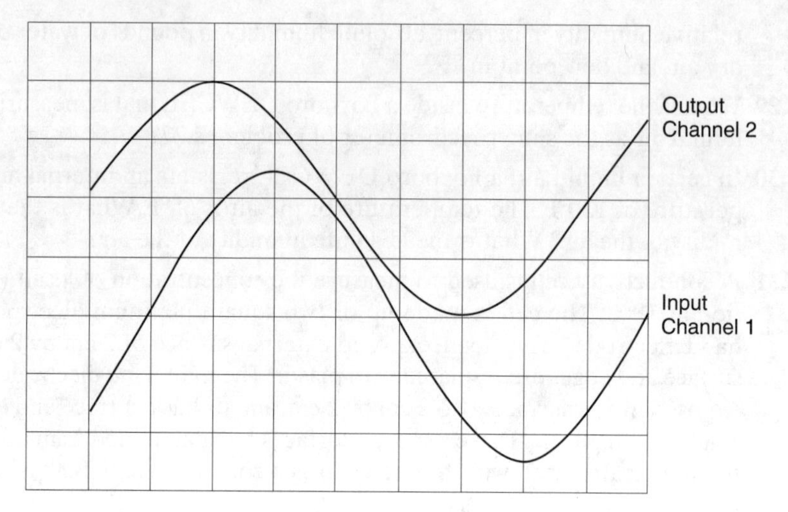

FIGURE 2.23
Figure for problem 2.36

tolerance resistor. This 4–20 ma signal drove a control valve that produced approximately 0–90 GPM of flow through the flow sensor. What is the phase shift of the output relative to the input in trigonometric degrees? What is the gain in ma/ma of the output relative to the input?

2.37 Steady-State Calibration of a Pressure Transmitter

Objective: To perform a steady-state "as-found" and "as-calibrated" calibration on a pressure transmitter.

Equipment:
- Electronic pressure transmitter (span approximately 10 psig) with 24 V DC power supply and accurate 4–20 ma reference milliameter.
- Reference pressure test gauge (0–15 psig) and 20-psig air supply

Procedure: Set up a pressure transmitter for calibration using an air supply with pressure regulating valve and a reference test gauge for the standard pressure input. Connect a reference milliameter with 24 V power supply on the output to display the actual output milliampere signal. Take "as-found" and later "as-calibrated" readings of standard input and actual output values at approximately 0, 25, 50, 75, and 100% of the transmitter span. Record the values on the instrument calibration sheet and graph them. For the "as-calibrated" readings, carefully increase the input pressure from 0% up to the reading near 25%. Do not decrease the pressure at any time. Then increase to the next reading, and continue increasing till about 100% is achieved. Then take readings while only decreasing the pressure toward 0%. Ensure these readings are plotted on the output error or deviation graph to amplify the error and to show the hysteresis effect in the error if it exists.

Conclusions: Include your graphs in your shop assignment report. What is the maximum error in psi and percent of span? How repeatable are your results in psi and percent of span? What is the value of the zero error in psi and percent of span?

2.38 Steady-State Calibration of a Link/Lever Pressure Recorder

Objective: To perform a steady-state "as-found" and "as-calibrated" calibration on a pressure recorder.

Equipment:
- Link/lever pressure recorder (span approximately 10 psig)
- Reference pressure test gauge (0–15 psig)
- Adjustable air supply from 0–20 psig

Procedure: Set up a pressure recorder for calibration using a pneumatic air supply with pressure regulating valve and a reference test gauge for the standard pressure input. Ask your instructor to ensure that the recorder is slightly off the ideal calibration. Take "as-found" readings of standard input and actual output values (read from the recorder pen or pointer) at approximately 0, 25, 50, 75, and 100% of the recorder span. Record the values on the instrument calibration certificate and graph them on the calibration graph only, and reserve the deviation graph for "as-found" results.

Ask your instructor to identify the zero adjustment, the multiplication adjustment, and the angularity adjustment. Minimize your *angularity error* by applying midspan pressure, and then adjust the length of the link connecting the input sensor lever to the output pen lever so that the angle between the lever, from the input to the link, is as close as possible to 90 degrees. Also try to achieve 90 degrees between the pen lever and the link at the midspan pressure. Then adjust the zero adjustment so that the pen points at midscale. Set the pressure at maximum value. If the pen does not read correctly, adjust the multiplication adjustment to position the pen about halfway toward the correct reading. Next set the pressure to the minimum value. If the pen does not read correctly, adjust the zero adjustment to position the pen at the correct reading. Repeat the adjustments for maximum and minimum pressures until you are satisfied that you cannot improve the calibration. For the "as-calibrated" readings, carefully increase the input pressure from 0% up to the reading near 25%. Do not decrease the pressure at any time, then increase by 25% to the next reading, and continue increasing till about 100% is achieved. Then take readings in decreasing steps of 25%, while only decreasing the pressure toward 0%. Ensure these readings are plotted on the same calibration graph with the "as-found" data to show how you improved the instrument. Then plot the "as-calibrated" data on the deviation graph.

Conclusions: Include your certificate with your graphs in your lab assignment report. What is the maximum error in psi and percent of span? How repeatable are your results in psi and percent of span? What is the value of the zero error in psi and percent of span?

2.39 Calibration of a Resistance Temperature Detector

Objective: To prepare a calibration certificate for an RTD (Resistance Temperature Detector).

Equipment:
- Platinum RTD 100 Ω at 0°C
- Accurate ohmmeter, reading to 0.1 Ω, compensated for lead resistance
- Reference thermometer
- Thermal flask
- Kettle for boiling water
- Ice cubes for making ice water

Procedure: Ensure ohmmeter reads 100 Ω for RTD after stirring RTD in ice water for at least five minutes. Obtain four precise readings of RTD resistance at ice water temperature, boiling water temperature, and a couple of intermediate temperature points using a thermal flask containing the RTD and a circular thermometer in a mixture of hot and cold water. Identify the RTD. On the certificate record the temperature of the liquid in column A for "standard input signal" and the measured resistance in column D for "actual output signal." The percent input column is the same as the "standard input signal" (0% input is 0°C, 100% input is 100°C). For column C, "the desired output signal," use the value of resistance in Appendix F corresponding to the temperature in column A. Calculate the "percent output signal" using the following equation:

$$\text{Percent output} = \frac{(\text{RTD } \Omega - \text{Appendix F } \Omega \text{ at } 0°\text{C}) \times 100\%}{(\text{Appendix F } \Omega \text{ at } 100°\text{C} - \text{Appendix F } \Omega \text{ at } 0°\text{C})}$$

Calculate the "percent output error," in this case nonlinearity, from the equation:

$$\text{percent output error} = \text{percent output signal} - \text{percent input signal}$$

2.40 Speed of Response of an RTD

Objective: To prepare a graph showing the speed of response of an RTD (Resistance Temperature Detector) and to read off the time constant from the graph.

Equipment:
- 100 Ω at 0°C RTD in a heavy brass thermowell
- Digital ohmmeter
- Reference thermometer
- Thermal flask
- Kettle for boiling water
- Ice cubes for making ice water
- Watch reading to nearest second

Fill the kettle with water and raise the water temperature to the boiling point. While the water is heating, make ice water in the flask. Connect the RTD to the ohmmeter and measure the resistance at 0°C in the ice water. When the RTD is at 0°C, plunge it into the boiling water and try to read the resistance every three seconds until it is no longer changing. Then plunge it back into the ice water and repeat the measurement without stirring the water. Repeat the exercise, but this time stir the RTD rapidly in the ice water.

Conclusions: Plot two graphs. The first graph should show the rising and lowering resistance without stirring, and the second graph should show the rising and lowering resistance with stirring. What is the time constant from your graph when the temperature is rising? What is the time constant from your graph when the temperature is falling in ice water without stirring? With stirring? Is there any difference, and if so, why? The time constant is the time it would take the resistance to rise or fall 63.2% of the total change in resistance from start (0°C on rising) to finish (100°C on rising).

2.41 Construction and Operation of a Test Thermocouple

Objective: To construct a test thermocouple from thermocouple extension wire and to operate it for temperature measurement.

Equipment:
- 4 ft of thermocouple wire
- Thermocouple calibrator or millivoltmeter to read to 0.1 mV
- Room thermometer

Procedure: Make a measuring junction and a reference junction at the two ends of the wire. In the middle of the wire, separate one of the conductors so that you can measure the voltage appearing at its two opened ends. Find out the type of thermocouple that you have made from the insulation color and obtain a table for its values. (If type J or K, use Appendix F.) Specify how you established the type of thermocouple and specify where the table comes from. Place the reference junction in ice water and the measuring junction in boiling water. What voltage do you measure (include polarity in your measurement)? How does it compare with the table? What voltage do you measure when the reference junction is allowed to reach room temperature while the measuring junction remains in the boiling water? Measure room temperature and calculate the voltage that you should measure. How does it compare with the table? What voltage do you measure when you allow both the measuring junction and reference junction to reach room temperature? Is that what you expect? Why? Measure the temperature of a soldering iron hot tip with your thermocouple. Show how you calculated this temperature. Draw a schematic of your thermocouple, include the names of the two metals and their polarity.

2.42 Calibration of a Pressure Switch

Objective: To calibrate a blind pressure switch so that it will operate electric circuit A if its applied pressure rises above the setting, or conversely, operate electric circuit B if the pressure falls below the setting.

Equipment:
- Blind (without gauge) low pressure switch (e.g., Dwyer Magnehelic, 0–100 in. of water) with normally open and closed contacts
- Ohmmeter
- Shop standard pressure gauge or manometer that covers the range of the switch

Connect the gauge and switch to an adjustable air supply. Connect the ohmmeter to the normally open contacts and raise the pressure until the contacts close.

Have your instructor select a suitable pressure for switch operation. Calibrate the switch to operate at that pressure by setting the air supply to that pressure and then adjusting the switch actuation point until the ohmmeter shows that it switches. Check and demonstrate to the instructor that it switches at the selected pressure. Check and demonstrate that the normally closed contacts also switch at that pressure. Find the dead band in pressure units between the pressure at which the contacts just open and the pressure at which they just reclose when the pressure goes in the opposite direction. Record these values for your report.

2.43 Calibration of a Magnetic Flow Transmitter Using a Tank

Objective: To calibrate a magnetic flowmeter at five flow rates from 20–100% rated flow by collecting a measured quantity of water in a tank during a measured period of time.

Procedure: Use a piped magnetic flowmeter emptying into a calibrated container. Establish the flow rate range (e.g., 0–10 GPM) of the magnetic flowmeter and its output signal (e.g., 4–20 ma DC). Set the flow to 20% of this range. With the flow running smoothly into the tank, note the level reading and start timing the run. Read the flowmeter output signal value every three to five seconds and record this value in a list. When the tank level has increased for at least five minutes, stop timing and note the level value. Record the time spent between the start and finish. Calculate the rate of change of volume (e.g., Y GPM) in the tank from the start of timing to the end of timing. For the calibration certificate, record this as the standard input signal. Record the average of all the output signal values for this 20% run as the actual output signal. Repeat for four more runs at approximately 40%, 60%, 80%, and 100% of maximum flow rate.

Prepare an "as found" calibration certificate and graph. If you have an error exceeding 4% of full scale, ask your teacher how to recalibrate the flowmeter.

Conclusions: How accurately can you calibrate this flowmeter? In your estimate of flow accuracy, include estimates for the accuracy of reading volume and time and the effects of these errors on your overall accuracy.

2.44 Calibration of a Liquid Level D/P Transmitter

Objective: To calibrate a D/P (differential pressure) transmitter for measurement of liquid water level in an open tank by directly measuring the liquid pressure at the bottom of the tank and comparing it against a sight gauge on the outside of the tank.

Equipment:
- Piped D/P transmitter on an open tank with an external level sight gauge

Procedure: Have your instructor assign a zero and span in units of level as measured by the sight gauge. Establish first the "as-found" calibration of the transmitter by using the sight gauge as the reference. If the error exceeds 1% of full scale, then recalibrate the D/P transmitter by adjusting its zero and span. Prepare a calibration certificate and graphs. Plot the "as-found" results on the "as-calibrated" calibration graph (not on the deviation graph) to show how you improved the instrument.

2.45 Calibration of an Ultrasonic Liquid Level Sensor

Objective: To calibrate an ultrasonic level transmitter by comparing it to a sight gauge on the outside of the tank.

Equipment:
- Piped ultrasonic level transmitter on the top of a tank with a sight gauge on the outside of the tank

Procedure: Have your instructor assign a zero and span in units of level as measured by the sight gauge. Establish an "as-found" calibration using the sight gauge as the reference. If the error exceeds 2% of full scale, then recalibrate the transmitter by adjusting its zero and span. Prepare a calibration certificate and graph. Plot the "as-found" results on the "as-calibrated" calibration graph (not on the deviation graph) to show how you improved the instrument.

2.46 Humidity Measurements: Psychrometer vs. DEWCEL

Objective: To compare humidity measurements of three different air locations using a sling psychrometer against humidity measurements using a lithium chloride DEWCEL.

Equipment:
- Sling psychrometer
- Foxboro DEWCEL with thermometer (0–200°F, readable to 0.2°F preferred)

Procedure: Be careful that you do not hit anything when you twirl the psychrometer. Allow about five minutes for the DEWCEL to stabilize after turning it on.

In a table record readings of wet and dry bulb temperatures and the DEWCEL temperature, first in the lab, then in the hall or stairwell, and finally outdoors. Use the psychrometric chart (Figure 2.16) to translate the wet and dry bulb temperatures into absolute humidity, relative humidity, and dew point (or saturation) temperature. Record these values with correct units in your table. Use the Foxboro DEWCEL operating characteristic chart (Figure 2.17) to translate the DEWCEL (internal or ambient) temperature to dew point temperature. Record this value in your table.

Conclusions: Compare the error in the dew point between the psychrometer and the DEWCEL. Why do you think this discrepancy occurs?

2.47 Electrical Conductivity in Liquids

Objective: To check the calibration of an electrolytic conductivity cell with distilled water and with a known concentration of salt (NaCl), and to measure the conductivity of tap water.

Equipment:
- Conductivity cell with manual and meter
- 1000 ml beaker
- 2 l of distilled water
- 30 mg of pure salt (NaCl)
- Thermometer

Procedure: Establish the cell constant from the manufacturer's data sheet. Ensure the meter is reading by first placing the cell in tap water. Read the the tap water conductivity and record it. First, rinse the thermometer and the conductivity cell with the distilled water. Then, with about 400 ml of distilled water, measure and record the conductivity. It should be zero. Replace this distilled water with 1000 ml of fresh distilled water and record the value of conductivity on the indicator and the water temperature. Add the 30 mg of pure NaCl to the 1000 ml of distilled water and record the value on the indicator. Multiply this reading by the cell constant. At 25°C the actual value should be 64 μsiemens/cm, and at 18°C it should show 60 μsiemens/cm.

Conclusions: Compare your reading of known salt solution conductivity with the expected values. Is there a difference greater than 10%? If so, why do you suspect this discrepancy occurred? Does the tap water have a concentration of salt greater than 0.003%?

2.48 Electrical Instrument Transformer Applications

Objective: To measure line voltage, line current, load power, and load power factor through the use of instrument transformers.

Equipment:
- One resistive load bank of 9 (0.5 A) switchable resistors, each rated 120 V, 60 Hz
- One inductive or capacitive load bank of 9 (0.5 A) switchable inductors or capacitors, each rated 120 V, 60 Hz
- Potential transformer 120/60 V
- Current transformer 10/5 A
- 0–100 V 60 Hz voltmeter
- 0–5 A 60 Hz ammeter
- Wattmeter 0–120 V, 0–5 A, 60 Hz

Procedure: Connect the two load banks in parallel. Although the voltage is only 120 V, for this assignment it is assumed that this is a high-voltage, high-current circuit. Draw a complete elementary diagram showing the load banks in parallel, and incorporate a potential transformer and a current transformer. Show on your drawing the meters on the secondaries of the transformers. The meters are to include one voltmeter, one ammeter, and one wattmeter. Show on your drawing the polarities of the instrument transformers and of the wattmeter coils. Also show the transformer ratios and the instrument ranges. Connect the components using your diagram and switch on the load and for nine steps (one R and one X per step) record all meter readings. Make a table and for each step show the number of Rs and Xs, the amp reading, the volt reading, and the watt reading. In the table make columns for calculated line amps, line volts, load power, and load power factor.

Conclusions: Calculate the load power and load power factor based on voltage and load bank ratings only and compare these calculations to the values found above. Is there a discrepancy? What do you think causes it?

2.49 Dynamic Calibration of a Pressure Transmitter

Objective: To measure the dynamic "as-found" relationship of the output signal of a pressure transmitter to the input signal over a wide range of frequencies.

Equipment:
- The pressure transmitter of problem 2.37
- Sine wave generator with a low frequency of at least 0.005 Hz
- Current-to-pneumatic converter
- Two-channel oscilloscope or oscillograph
- Two 250 Ω load resistors
- 20 psig air supply
- 24 V DC power supply
- 0–15 psig reference pressure gauge

Procedure: Study section 2.11 in this chapter. Connect the sine wave generator to drive the current-to-pneumatic converter whose pneumatic signal is the input to the pressure transmitter. Include the pressure gauge to display the pneumatic signal when it is excited by the sine wave. Connect a two-channel oscilloscope or oscillograph so that one channel measures the current (actually voltage drop across a load resistor) to the current-to-pneumatic converter and the other measures the current (actually voltage drop across a load resistor) from the pressure transmitter. Draw a schematic diagram showing your electrical and pneumatic system. Start the sine wave generator so that a pneumatic signal of about 9 psig is the average value and the sine wave amplitude is about 3 or 4 psi. Start with a frequency of about .01 Hz and record the values of frequency, peak-to-peak input and output current, and phase shift in trigonometric degrees. Ensure that you have the correct relationship between input current and pressure. Double the frequency and repeat the readings. Continue doubling the frequency and taking readings until the output amplitude is 0.01 times its value at the first frequency or the phase shift is more than 200 degrees.

Conclusions: Calculate the gain values for each frequency for the pressure transmitter alone by dividing the output peak-to-peak milliampere value by the input peak-to-peak pressure value. What are the units of gain for the pressure transmitter? Plot the Bode plot for the pressure transmitter using log values for the frequency and for the gain and using linear values for phase shift. Include manufacturer and model number on the sheet, and include units on the ordinate and abscissa. Note the point where the gain is 0.7 times its initial low frequency value. The associated phase shift should be about –45 degrees, and the associated frequency should be close to the point where the gain curve starts to cut off the output from its initial low frequency values. Did you really have a good measurement of the sine wave of air pressure?

REFERENCES

American Society of Mechanical Engineers. 1971. *Fluid Meters, Their Theory and Application: Report of ASME Research Committee on Fluid Meters.* 6th ed. New York: American Society of Mechanical Engineers.

Anderson, N. A. 1980. *Instrumentation for Process Control.* 3rd ed. Radnor, PA: Chilton Company.

Bateson, R. 1996. *Introduction to Control System Technology.* 5th ed. Upper Saddle River, NJ: Prentice Hall.

Considine, D. M. 1985. *Process Instruments and Controls Handbook.* 3rd ed. New York: McGraw-Hill Book Company.

Kirk, F. W. and Rimboi, N. R. 1975. *Instrumentation.* 3rd ed. Chicago: American Technical Publishers, Inc.

Liptak, B. G. and Venczel, K. 1982. *Instrument Engineers' Handbook—Process Measurement.* Rev. ed. Radnor, PA: Chilton Company.

O'Higgins, P. J. 1966. *Basic Instrumentation Industrial Measurement.* New York: McGraw-Hill Book Company.

Tippens, P. E. 1978. *Applied Physics.* 2nd ed. New York: McGraw-Hill Book Company, Gregg Division.

3

PRESENTATION OF DATA

OBJECTIVES

When you complete this chapter you will be able to:

- Describe the presentation of a pressure gauge and a strip chart recorder
- Use an overview display to focus on an out-of-spec process variable
- Select a group display to adjust the set point or manual output of a control loop
- Interpret a point display
- Interpret a trend display
- Interpret an hourly average display
- Interpret an alarm summary display

The purpose of data presentation is to provide a human or a machine with values of related process variables over a period of time so that the status and trend of the process may be monitored continuously. The three major signals on the controller block of each control loop (see Figure 1.1) are presented to humans. They are the set point, the measuring sensor feedback, and the controller output signals.

The production of cement clinker in a cement kiln is an example emphasizing the purpose of data presentation. The operator will review many instruments as he controls the kiln, but a multipoint strip chart recorder presents him with the kiln temperature profile, using several sensors along the kiln, and with the kiln drive motor amperes, kiln draft pressure, and a few other variables. These variables appear as trend curves for the past three or four hours so that he can see the status of the process as it changes and, if necessary, adjust the fuel firing rate, kiln-turning speed, and air flow rate.

The presentation of data is defined as the translation of process variables into a format that conveniently displays for humans or presents for machines the changes of the process variables with time. In the cement kiln example, the strip chart recorder presents to the human operator several related cement kiln variables in graphical format for the most recent three or four hours. As an example of presentation to a machine, the 4–20 ma signal from a temperature sensor presents to a temperature controller a feedback signal corresponding to the temperature.

The Instrument Society of America standard, ISA-S5.1 Instrumentation Symbols and Identification, establishes a uniform means of designating instruments used for measurement and control. These symbols are shown in Appendix A. These symbols apply to process flow plans showing how instrumentation interacts with the process. The symbols are appropriate for use in the presentation of data to human operators. They are often used on computer displays showing the process flow plan to indicate the position in the process where the data comes from.

3.1 PRESSURE GAUGE

One of the simplest devices used to present data to a human operator of a process is a pressure gauge. About 1830, Eugene Bourdon invented the Bourdon tube pressure gauge. Before that the manometer was used to measure pressures less than 100 kPa.

The process fluid enters the pressure gauge to drive its indicator to the correct position corresponding to the pressure. The operator looks at the gauge to obtain a reading of pressure. This device is powered by the process. Because it contains the process fluid, it can have problems of safety, mounting, and location. If the process fluid is a hazardous one, or if it can freeze, then safety of personnel or safety of the process equipment must be considered. If the process tap is located high up on the equipment, then the head, due to the liquid in the downcoming line, can cause an error if the gauge is located near the ground. In most of these cases, a pressure transmitter would solve the problem, but a pressure gauge is simpler and less expensive. If a record of pressure is required, then the operator must write down the value that he reads from the gauge. Whereas a transmitter can transmit the value over a long distance to a computer for recording along with many other associated variables. Good transmitters have better accuracy than good pressure gauges. The time that a variable is recorded by a computer is very precise compared to manual records, and the frequency of reading variables by a computer can be much greater than a human can do. A pressure gauge presents valuable data to a human operator, but it does not achieve the capability required in most current applications.

3.2 MULTIPOINT STRIP CHART RECORDER

The presentation of data to a human is exemplified by a multipoint strip chart recorder as shown in Figure 3.1. The characteristics of presentation are:

- Format—in this case trend curves of dots drawn by a pen on a paper graph for process variables *A* and *B*.
- Accuracy—depends on pen calibration and resolution plus the calibration of the sensors and the shielding of the transmission medium from interference.
- Timing—the vertical coordinate shows the time when each dot occurred in hours and minutes.

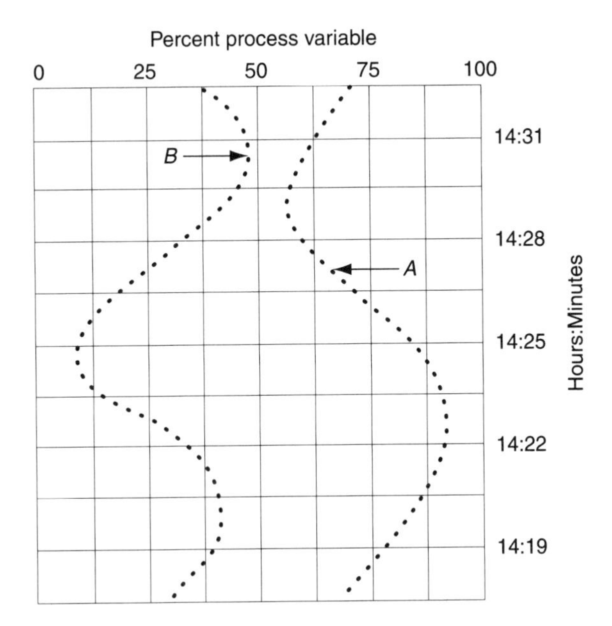

FIGURE 3.1
Multipoint strip chart

- Frequency—the frequency of updating the curve is shown by the time between the dots. In Figure 3.1 this is one dot per 18 seconds. If a computer is used to acquire the data, then the presentation on a video display to a human may very well take a similar form to Figure 3.1.

3.3 SINGLE-LOOP FLOW RECORDER

A single-loop recorder associated with a controller is often used for process monitoring and for tuning the loop to achieve some criterion of loop response. Such a loop includes presentation to a human—the recorder does this—and presentation to a machine—the electrical 4–20 ma signal from the sensor to the controller does this. Such a recorder has a pen to show the flow rate and usually has additional pens to show the set point and the output signal to the process adjusting device. Figure 3.2 shows such a recorder displaying the responses of the flow and the output signal (sent to the control valve) to a step change in set point.

3.4 COMPUTER ACQUISITION SYSTEM

Computers driving video displays or operating as data analyzers have revolutionized the presentation of data to humans and to machines. For humans any type of picture can be drawn on the display, and it can be updated almost instantly. This is a vast improvement over a pointer above a scale or pen drawing a single line on paper. For machines the data can be thoroughly refined for special kinds of automatic control, including the latest kind of "expert system."

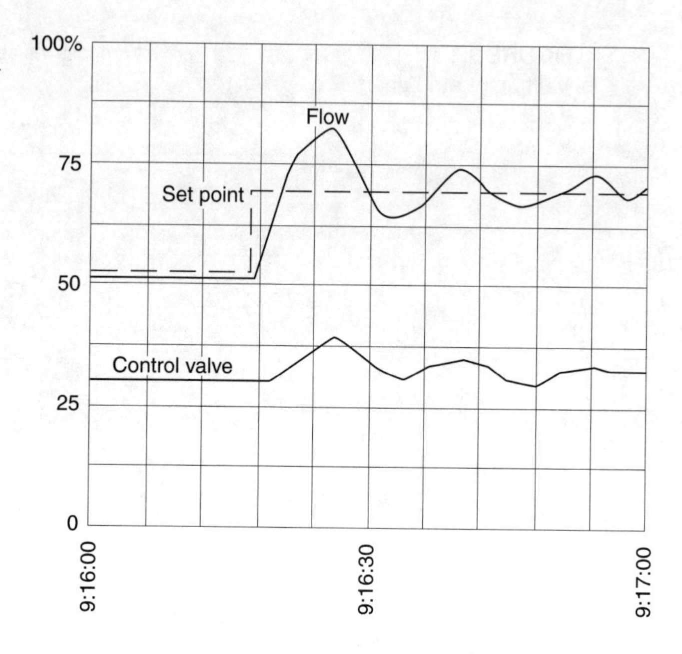

FIGURE 3.2
Single-loop recorder

To present information to humans, there is a hierarchy of displays provided on most of the commercially available systems. These displays start with the "big picture" or overview, showing a little bit, whether the variable is inside or outside alarm limits, about all the measurements on a process. The operator can then zoom in on part of the process and see only a few related variables—but a lot more information about each of them. Then the operator can zoom in further to obtain details about one or two variables. For each level of zoom there is a standard display format. For example, if we examine the Honeywell TDC-3000 distributed control system, we see the overview display (Figure 3.3), the group display (Figure 3.4), and the point display (Figure 3.6). In addition the custom graphics display allows the user to generate whatever kind of display the user wants (such as Figure 3.5). With the use of color, these powerful diagrams help the operator to absorb much more process information in much less time than in the past.

Other types of displays or printouts are common to computer acquisition systems. These may include the trend display (Figure 3.7), the hourly average report (Figure 3.8), the alarm summary display (Figure 3.9).

3.5 OVERVIEW DISPLAY

This "big picture" (Figure 3.3) shows the status of up to 36 groups, with each group having up to eight variables, for a maximum of 288 variables. Each group is identified with a number and a short title. As the operator becomes familiar with each group, the operator recognizes the position of each variable and sees whether the variable is within limits or if it is in manual control mode (an M appears on the group

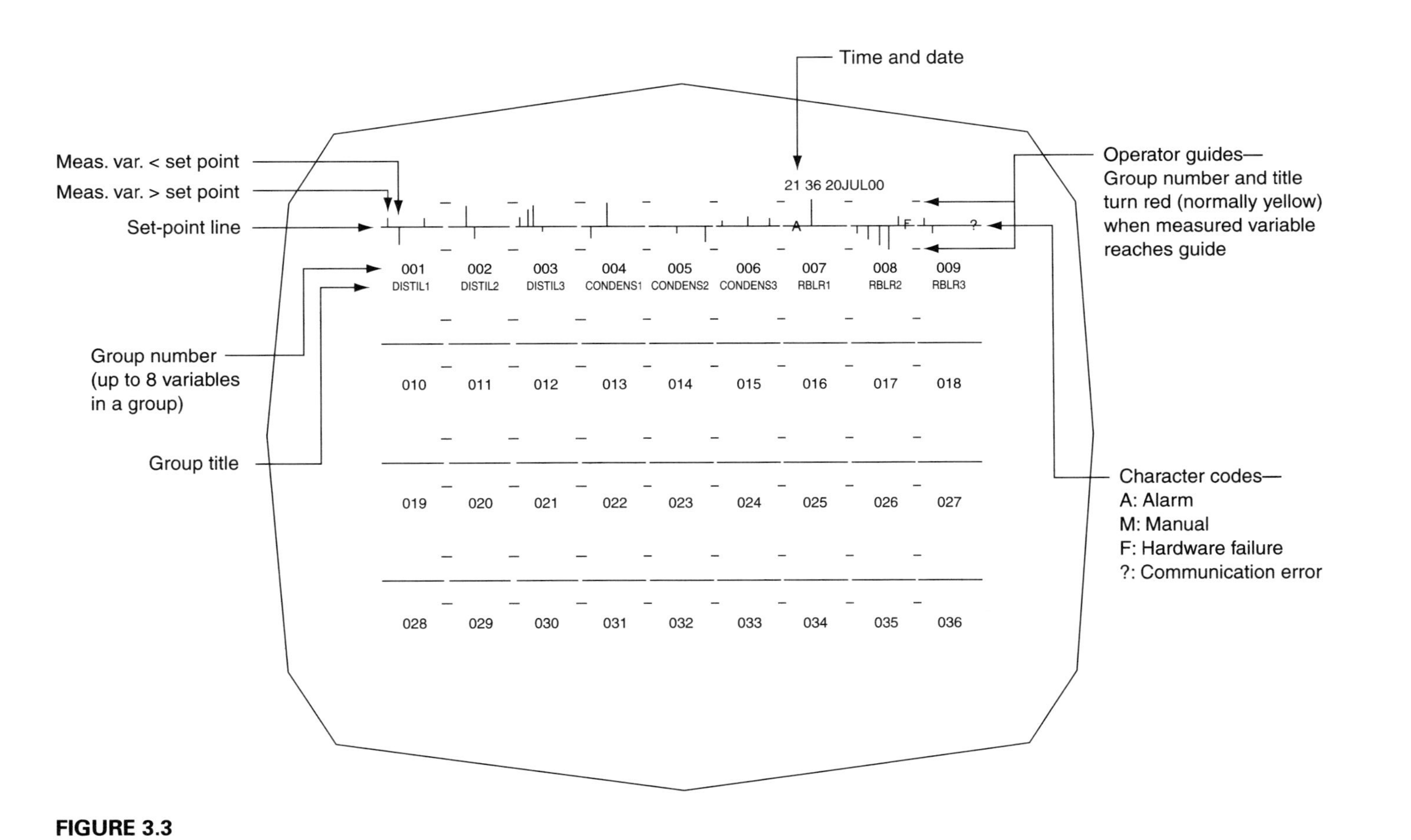

FIGURE 3.3
Overview display of the Honeywell TDC 3000

set-point line). Manual control mode is described in section 6.3. The horizontal set-point line is shown with a space separating each group. As any of the up to eight analog (continuous) process variables in a group deviates from its set point, a short vertical line appears starting from the set-point line. If the deviation is such that the process variable is less than the set point, then the vertical line is below the set-point line. When the deviation from set point of any process variable in a group shown on the overview reaches the operator guide line, the group number and title change from yellow to red. The guide line values are determined from the overview index configured on the point display (Figure 3.6) for that process variable. If an alarm value, separately set from the overview index, is reached, then an A appears at the process variable position on the group set-point line. Digital on/off points do not show deviation, but they do show alarms, if configured to show them.

The first 36 groups are included in the overview, but many more groups may be specified in a system.

3.6 GROUP DISPLAY

For any analog point on the group display, the operator can change the set-point value, the control mode (manual, auto, cascade), and the controller output value when in manual mode. He does this by selecting one of the eight points on the group display using one of eight buttons. This places an X above the loop or tag name for that point on the display. He may then use the special set-point buttons to enter a numerical value or to increase or decrease the set point. In a similar way he may change the output of the loop. For digital points the operator uses special state change buttons to latch or pulse the selected digital point on or off.

The group display in Figure 3.4 shows seven analog variables and one digital (Run–Stop) variable. The group number, title, time, and date are shown. For each loop the measured variable or process variable (PV) is shown as a vertical bar graph and also numerically (108 on leftmost loop). The set point is shown as a crossbar and also numerically (105), and the output percentage is shown as a narrow vertical bar and also numerically (42.1). An alarm, the low process variable alarm (LP), is shown on the fourth loop only. The loop name (F1 on the leftmost loop) of the variable is shown along with engineering units, cc/m (cubic centimeters per minute) and abbreviated process descriptor (FEED). The current control mode (A for automatic) is shown. The leftmost loop, F1, has been selected (see the point selector X) by the operator for changes, such as set point changes or output changes or control mode changes. At the bottom the unabbreviated process descriptor is shown for the selected loop.

The engineer configures each group to include those continuous and digital points that are closely related on the process. Very important points may be repeated in several groups. The basic building blocks of a process control system are the analog loop and the digital point. For example, an instrumentation flow plan is prepared by the control engineer with the process engineer, according to the Instrument Society of America Standard S5.1 (see Appendix A). The example in Figure 3.5 transfers fluid from tank A to tank B. Continuous flow adjustment is required from

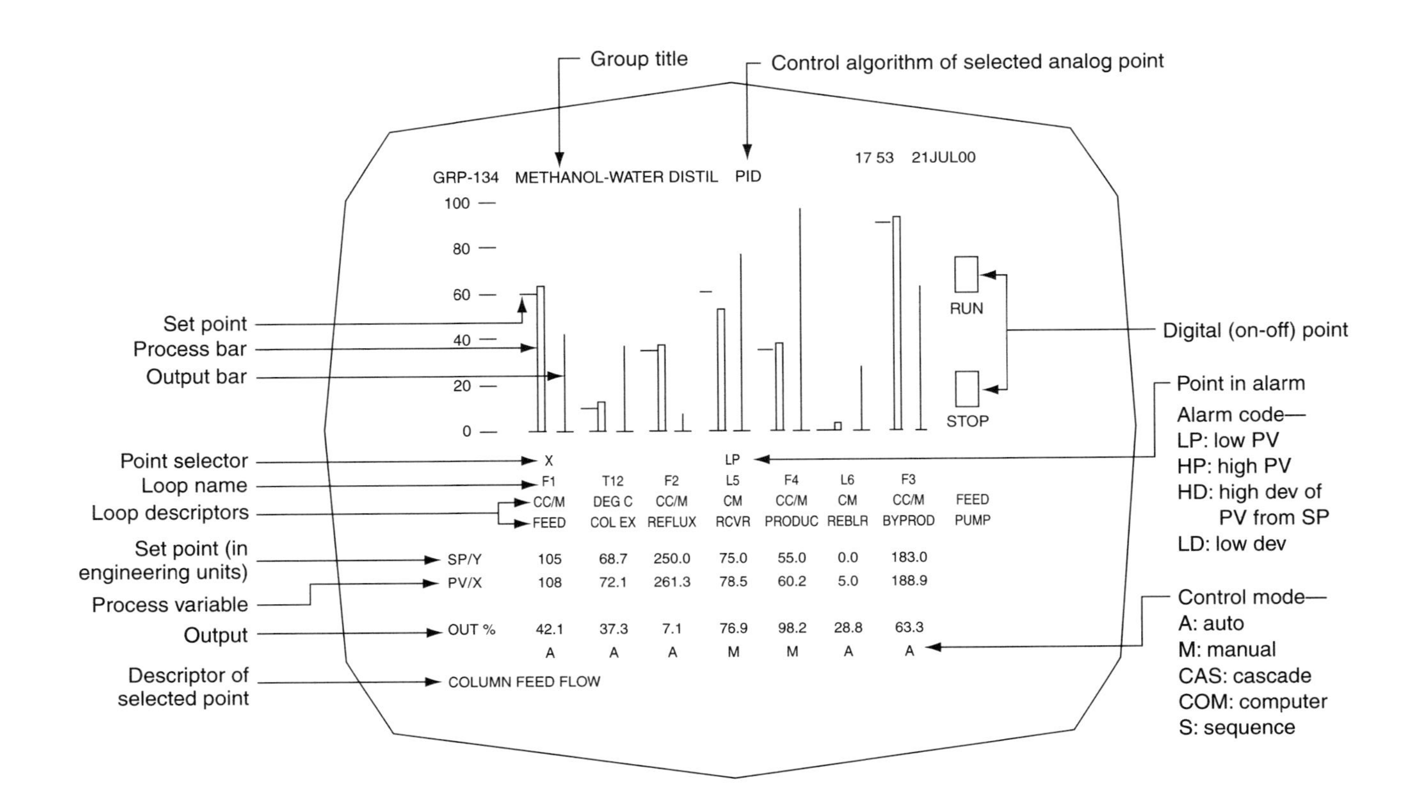

FIGURE 3.4
Group display of the Honeywell TDC 3000

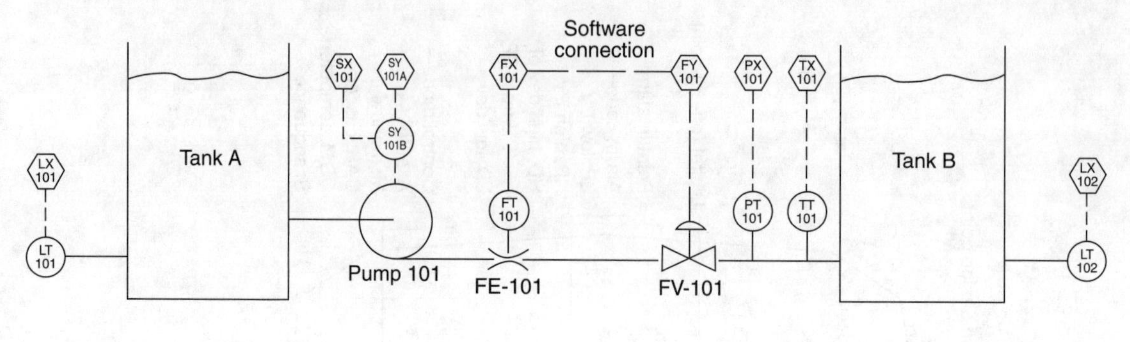

FIGURE 3.5
Instrumentation flow plan

maximum flow down to 20% of maximum. Temperature and pressure measurements are required. And a pump driven by a single-speed electric motor is specified for this process. Therefore, required for this process are:

- A closed flow loop
- Measurement-only semi-loops for temperature, pressure, level in tank A, and level in tank B
- A motor on/off digital point

Obviously, these six points are specified on a single group display. The operator can call up this display at any time. The operator can select the pump point to start or stop the process—or the flow control loop to adjust the rate of flow over any feasible range. While the operator makes a change, the operator can see the effects on all six variables as they happen. However, on regular displays, the group values are updated by the computer once every five seconds, which may be a bit slow. If fast update is requested, then it is updated once every ⅔ second. Personnel cannot change the points configured into each group after system startup unless they perform a key-lock security procedure. Normally the groups for the complete system are configured by selecting the points that are to make up each group. Then these point or tag names are combined offline in libraries for each group number with the group name. See section 3.7 for each point configuration.

3.7 POINT DISPLAY

The detail associated with one process variable is presented on the point display in Figure 3.6. The information on the group display is duplicated for the variable in the point display with controller tuning constants (described in Chapter 6), alarm settings, and the loop hardware configuration, plus other parameters. To the right of the

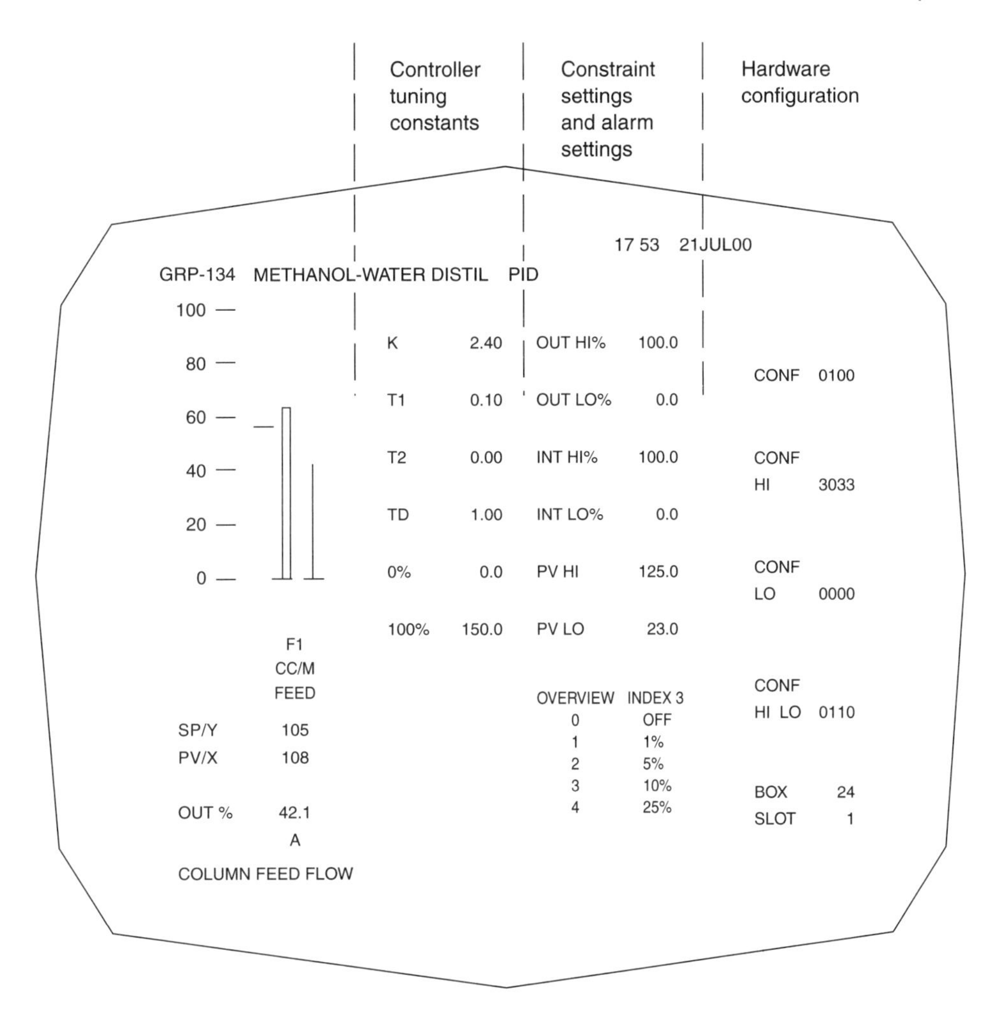

FIGURE 3.6
Point display of the Honeywell TDC 3000

bar graphs are values entered by the control engineer for the loop controller tuning constants and range. These include gain, K (0.1–100); integral time, T_1 (0.02–91 minutes/repeat); derivative time, T_2 (0.2–99 minutes); a *PV* (process variable) filter time constant, *TD* (0.006–17 minutes); and the 0% and 100% range values in engineering units. Further to the right on Figure 3.6 are constraint settings for the output signal, integral action, and settings for process variable *HI* and *LO* alarms and for the overview alarms.

This figure shows a flow control loop similar to the loop F101 in Figure 3.5. The operator may change the set point, the control mode, and the output from this display in the same way as the operator changes them from the group display (see section 3.6).

However, to change the controller tuning constants or the constraint or alarm settings, the operator must be authorized to use the security keylock prior to changing these values. For example, suppose the operator wants the controller tuning to be 80% proportional band, 0.07 repeats/second, no derivative action, one second filter time constant. He turns the keylock to the data entry position and moves the cursor on the display to 2.40 for K (the gain) and enters 1.25, which is the gain corresponding to a proportional band of 80%. Then the operator moves the cursor to the 0.10 for T_1 and enters 0.238, which is the integral time in minutes/repeat corresponding to 0.07 repeats/second. Then the operator moves the cursor to the 1.0 for TD and enters 0.0167 minutes. If the operator wants to change the overview guide lines from 10% of the span of the process variable to 5%, then the operator moves the cursor to the 3 (at the overview index) and changes it to 2. (The operator does not require the keylock security authorization to change the settings for the overview guide lines.)

3.8 TREND DISPLAY

The trend display (Figure 3.7) can be selected for 60 averages each of 20 seconds duration, or one minute, three minutes, six minutes, or 12 minutes. As each new average comes in and updates the display at the right, the oldest bar disappears from the left.

The trend display is similar to the single-loop flow recorder of Figure 3.2. With the group display showing and the variable selected with an X, the operator simply presses the special trend button followed by the number to get the average duration. The trend display then builds up with a bar for each average duration period. The TDC-3000 system has the facility to include two variables on the one trend display. The second variable is laid over the letters and numbers under the first trend.

3.9 HOURLY AVERAGE DISPLAY

The hourly average display (Figure 3.8) replaces the bar graphs on the group display with ten hours of averaged numerical values to form a shift report for the group that may be printed on the printer as well as shown on the display.

3.10 ALARM SUMMARY DISPLAY

The alarm summary display (Figure 3.9) is a special grouping of up to 100 alarm points, 20 points per page, showing high, low, and deviation status with tag and process descriptor with information on whether the alarm is flashing (not yet acknowledged by the operator), disabled, or steady (acknowledged, but still in alarm). When the point returns to a normal value, the alarm is simply removed from the display if it has been acknowledged. If more than 100 alarms have not been acknowledged or have

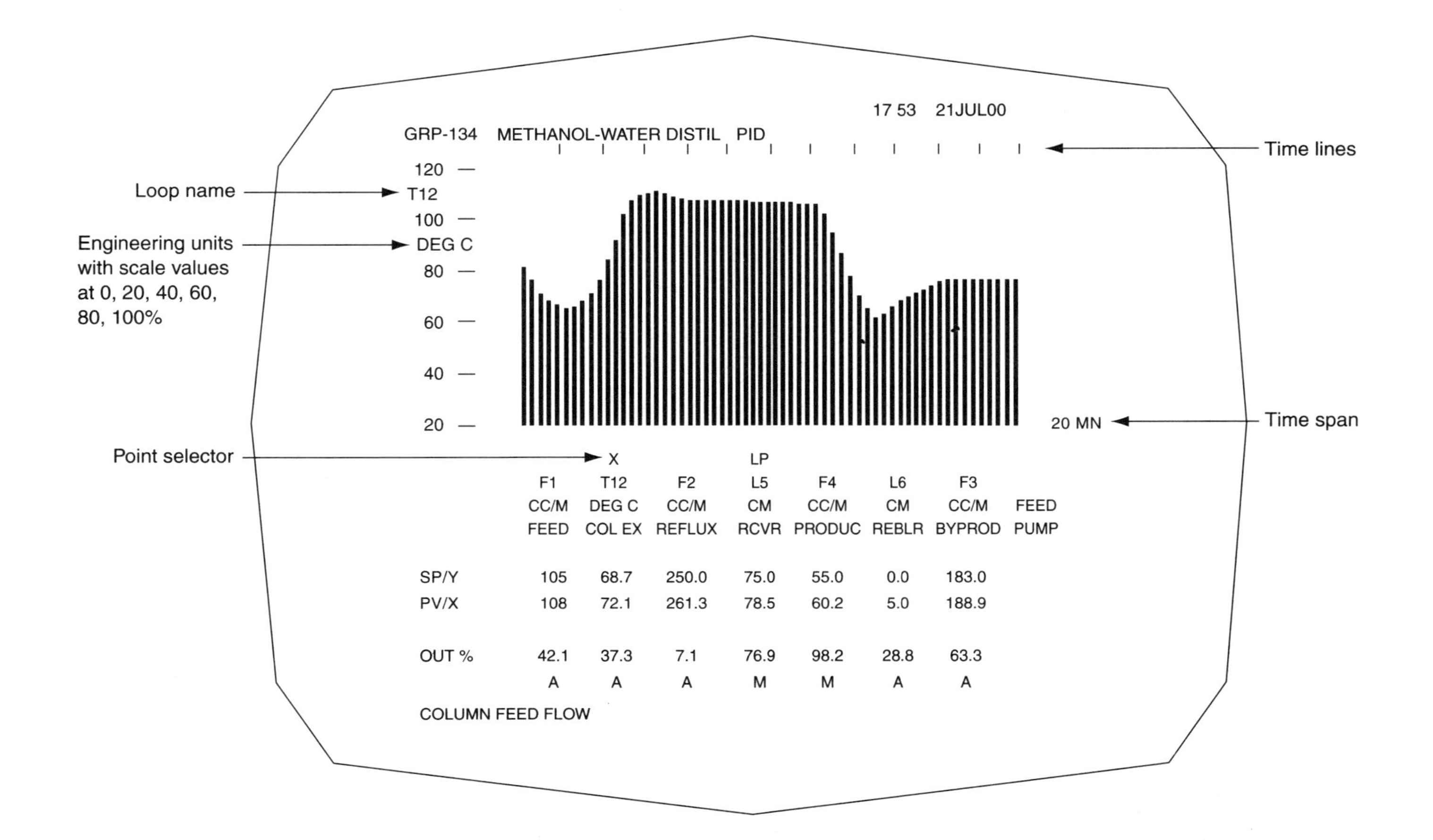

FIGURE 3.7
Trend display of the Honeywell TDC 3000

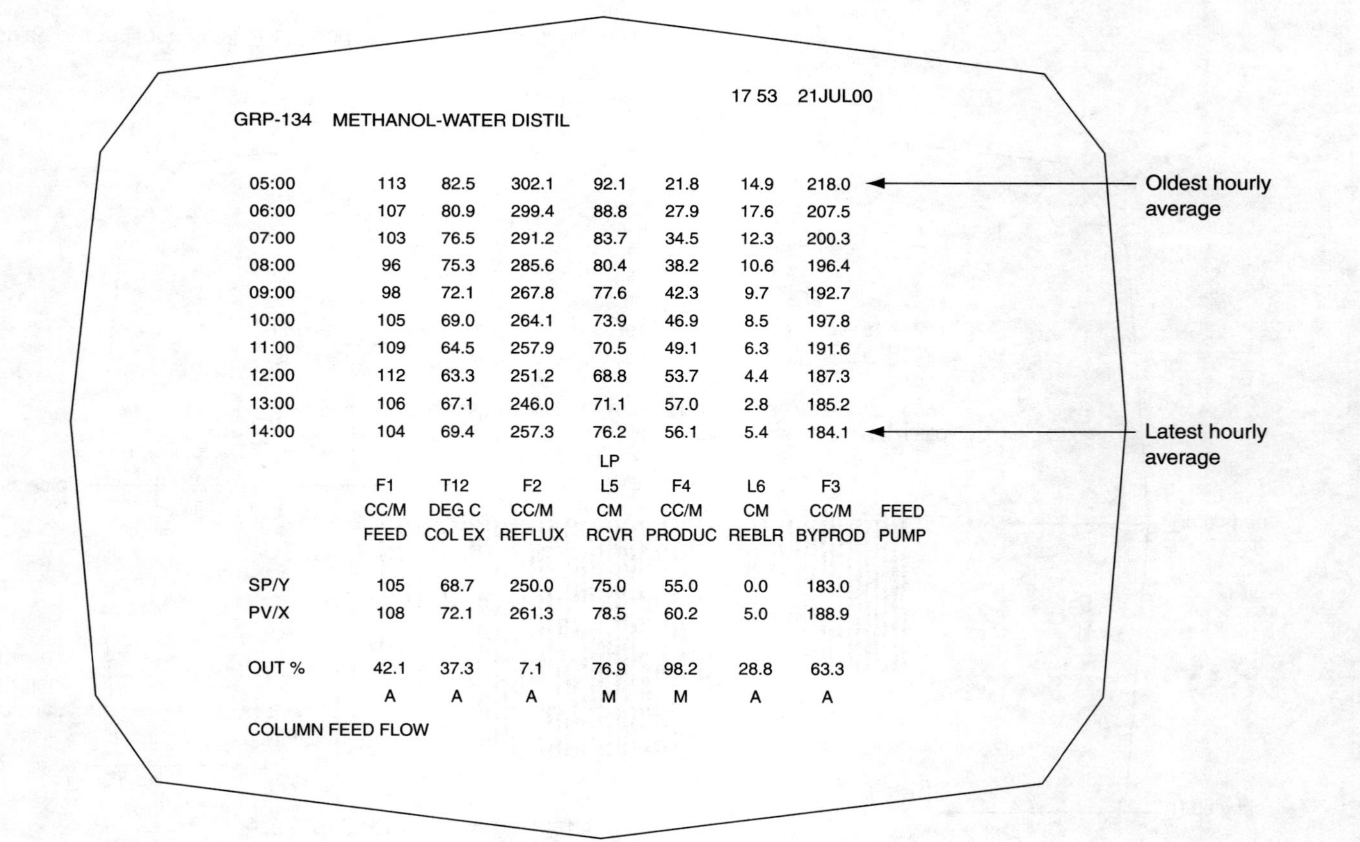

FIGURE 3.8
Hourly averages of the Honeywell TDC 3000

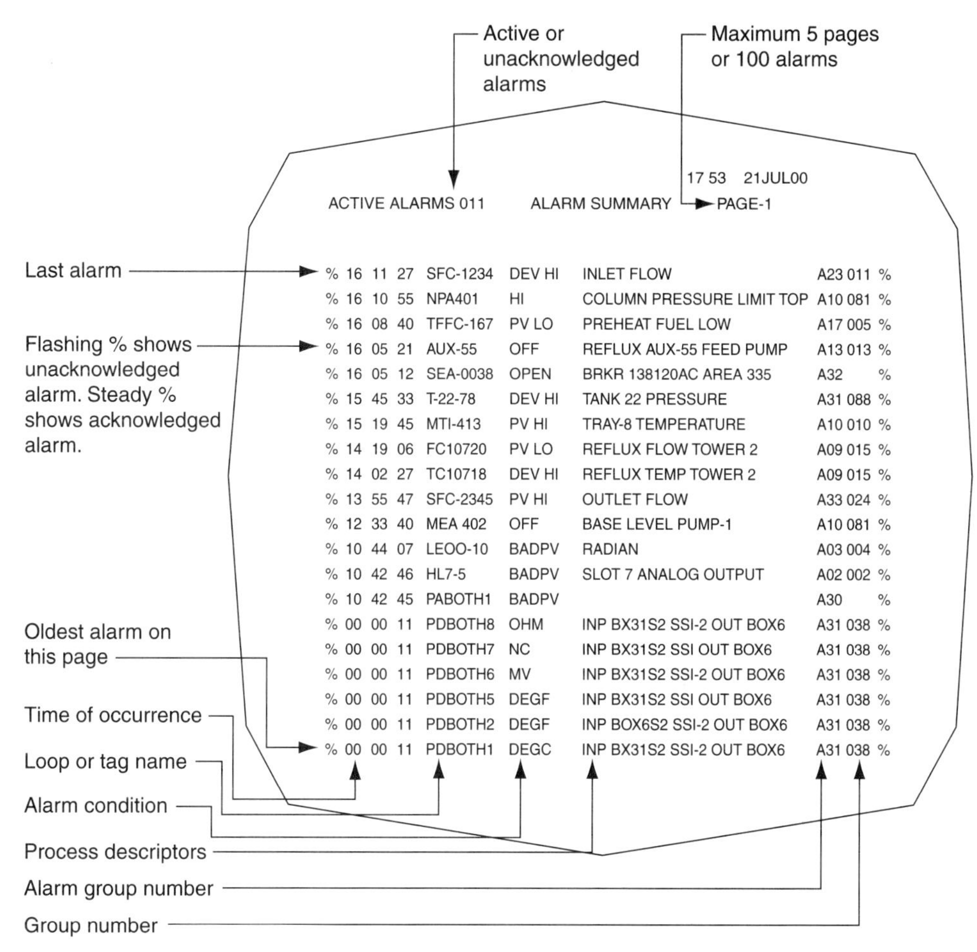

FIGURE 3.9
Alarm summary of the Honeywell TDC 3000

not returned to normal, then the message "EXCESS ALARMS" appears next to the number of active alarms, and the oldest alarms are dropped from the list, so that only 100 alarms remain in the list.

3.11 PRESENTATION SUMMARY

Relating the presentation displays to control loops, we can see that the three main signals (set point, sensor feedback, and controller output) at each controller block are available for the human operator to view. The operator can also adjust two of these signals (set point and controller output). The controller output signal sets the value of the process adjusting device. Therefore the operator sees the value of the manipulated process variable and the measured process variable.

In addition the operator can tune the controller using the point display, set alarms, and see the trends of the process by reviewing the trend displays, the hourly average display, and the alarm display.

PROBLEMS AND LAB ASSIGNMENTS

3.1 From Figure 3.1 estimate the frequency of the fastest sine wave (10 dots per cycle) that you could perceive well on the multipoint recorder.

3.2 For the second loop (starting from the left) of the group display of Figure 3.4, what is the value in engineering units of the set point and the process variable? What are the engineering units? For the fourth control loop shown, what is the value of the output signal and in what units? What is the control status or mode of this fourth loop? What ISA tag is used to designate this fourth loop?

3.3 From Figure 3.5 sketch a TDC-3000 group display showing typical values for six points, including one for the pump motor.

3.4 Sketch a TDC-3000 analog point display similar to Figure 3.6 using the following values: proportional band of 80%, integral action of 3.0 seconds/repeat, no derivative action, minimum digital filtering, 0–200 cc/min corresponding to 0% to 100% of range, output high alarm of 95%, output low alarm of 5%, integral action high alarm of 100%, integral action low alarm of 0%, measured variable alarm high of 178 cc/min, measured variable low alarm of 34 cc/min, overview index set for 25%, set point of 112 cc/min, output at 42.4%.

3.5 **Presentation of a Pressure Gauge vs. Pressure Transmitter plus Strip Chart Recorder**

Objective: To demonstrate the presentation made by a pressure gauge measuring water pressure in a tank and to compare it to a blind pressure transmitter with a strip chart recorder.

Equipment:

- Water tower between 4 ft and 20 ft high with a calibrated sight gauge
- Portable pressure gauge to measure tower pressure
- Pressure transmitter to measure tower pressure
- Strip chart recorder connected to transmitter

Procedure: Connect the portable pressure gauge with plastic tubing to the bottom of the water tower. Connect the transmitter also and ensure its recorder is recording the level by draining slightly and filling slightly the tower. With the tower full of water and the portable pressure gauge near the bottom of the tower, read the pressure in psig from the portable pressure gauge and the level in inches from the sight gauge. Mark these values on the recorder chart near the level pen value. Lift the portable pressure gauge up at least 30 in. above the base of the tower and repeat the readings. Lower the liquid level to 40 in. above the base and repeat the readings. Lower the level to 30 in. and repeat the readings.

Lower the level to 20 in. and repeat the readings. Draw a diagram using ISA symbols and show only the tower, the pressure gauge, and the transmitter and recorder.

Conclusions: As you changed the height of the portable pressure gauge, what happened to the value that it showed? As you changed the height of the liquid level, what happened to the values shown by each of the three level indicators? What essential requirements are needed to operate the transmitter and recorder, which are not needed for the simple pressure gauge? Suppose that you had a large tank with 10 ft of water in it and the bottom of the tank were 25 ft above the ground, what would a pressure gauge read if it were conveniently located 5 ft above the ground? What would a recorder (conveniently located 5 ft above the ground) read if it were connected to a transmitter located at the bottom of the tank? What device is necessary to keep the tank in operation, but allows the maintenance personnel to remove the gauge or transmitter to their instrument shop for repair? If the tank is outdoors, what is necessary for winter operation of the instruments when the temperature drops to –10°C?

3.6 Presentation of a Distributed Control System

Objective: To be introduced to a distributed control system, to become familiar with the overview display, group display, and detail display.

Equipment: ▪ Distributed control system (e.g., Honeywell TDC-3000)

Procedure: Follow the procedures in the computer control manual to load the operating system and the configuration into the system. Display the overview and print it out. Display the group display and print it out. Display a point display and print it out. Display the group again, select an analog point in the group display, put the point on manual control, adjust the output to 75%, and show the change in red on your printout. Now put the point in auto and change the set point to 15% and show the changes in red on your printout. Try to make a deviation alarm appear. If you make such an alarm appear, mark in the changes on your overview printout, group printout, and detail printout. Try to make other alarms appear. If they appear, describe what you did and what they look like. On the detail display, try to show the tuning trend display. Print it out. Also return to the group display and try to show the trend display for a selected point. Print it out. Then try to show the hourly averages display and then the alarm summary display. Print them out.

Conclusions: Combine your printouts into your report with your descriptions and comments.

4

DATA COMMUNICATIONS

OBJECTIVES

When you have completed this chapter you will be able to:

- Describe the media used to communicate control loop signals
- Describe the analog and digital formats of the control loop signals
- Describe the environments through which the control loop signals are designed to be able to communicate
- Describe the power supplies required by the control loop components
- Describe networks of control loop communication signals

4.1 INTRODUCTION

The purpose—and definition—of *data communications* is to send data frequently, reliably, safely, and in a useful format from the point of origin to the point of use. The transmission distance can be a few feet or many miles. The data could be required to be updated many times per second or once a shift. High/low status of a process variable may be satisfactory or a range of values with resolution of 1 part in 4096 (12 bits) could be required. The format of the data can be analog or digital, and the transmission medium can be electric wires, radio waves, optic fibers, or pneumatic tubing. A single pair of wires may be used, or a network of wires, optic fibers, switches, and converters could be interposed between the point of origin and the user. The transmission lines can run through a hazardous environment, and they can require a moderate amount of power to operate. Communications also implies responsiveness or a two-way interchange of information over the transmission line. Other terms used for data communications include *telemetry* and *transmission*.

For the control loop shown in Figure 1.1 on page 4, there are four data signals as follows:

Point of Origin	*Point of Use*
Process	Sensor
Sensor	Controller
Controller	Process adjusting device
Process adjusting device	Process

The signal from the sensor to the controller and the signal from the controller to the process adjusting device require communication as described in this chapter. These signals are typically transmitted over distances of several hundred meters. In addition, special care is required to introduce the process variable to the sensor. Imagine an incorrectly placed thermowell containing the RTD or thermocouple. It will not receive the correct temperature from the process. Or imagine the impulse lines for a pressure sensor that can be plugged up with fine solids. Again, incorrect pressures will be measured.

The communications environment introduces many problems requiring the selection of the appropriate medium, appropriate speed of transmission, appropriate format, and appropriate power supply. The selection is based upon the distance, safety requirements, potential interference, required frequency of updating, the cost of power, and simplicity.

The format of the data is usually analog, because that is the most common form available with most sensors. It is 4–20 ma DC for the electric wire medium. For pneumatic signals it is 3–15 psig or 20–100 kPa. Most sensors include hardware to translate from their basic signal, usually analog (e.g., resistance of an RTD), into one of these standard analog (e.g., 4–20 ma current) communication signals. The complete sensor, with this translation hardware, is usually called a *transmitter.* Recently, digital format is beginning to appear with many of the new smart transmitters, and it has added two-way data communication between the sensor and the computer or a handheld calibrator.

Most process measurement sensors cut off at frequencies above 10 Hz, which is equivalent to a maximum sampling rate of 40 or 50 times per second. Shannon's rule requires a minimum of 20 samples per second equal to 2×10 Hz. At the present this is an upper limit on the speed of updating the data from most sensors for industrial applications. A few special devices, usually for research purposes, do operate up to 10,000 Hz and may require sampling at 40,000 to 50,000 times per second.

Most electrically powered sensors use a power supply of 24–30 V DC, which can be made intrinsically safe from electric sparks in an explosive atmosphere. This usually is derived from the electric utility. In locations far from a utility outlet, however, solar power is coming into use on gas pipelines, railway signal towers, and other similar applications.

Networks introduce a great deal of complexity, because a device may communicate in both directions, sending and receiving messages from other devices, and because a message must not interfere with other messages from many other stations on the network.

4.2 MEDIA

Communications media include pneumatic tubing, optic fibers, radio waves, and electric wires.

Pneumatic tubing is less frequently used now for transmitting data than in the past. In any case it is not used if the distance is greater than a few hundred feet or if complex analysis and display is required. It is not used if digital formats are required. It is used for recording a single line on paper for each variable, or displaying a needle over a scale, or for simple proportional-plus-reset-plus-derivative control.

Optic fibers are beginning to be used for communication. But they are still expensive, and personnel are not very familiar with them. Special kits are required for terminating them.

Radio waves are used over longer distances of at least a few miles. However, radio waves can also be used in industrial plants to replace electric wires.*

Electric wires are the most common medium of communication today. They are simple, fairly inexpensive, and they can handle analog or digital formats over long distances at high speeds. They are subject to interference and can be hazardous in some dusty or gaseous atmospheres. For hazardous areas their circuits can be made intrinsically safe, or the wires can be placed inside explosion-proof conduits and boxes.

Pneumatic

Analog signals (3–15 psig or 20–100 kPa gauge) are transmitted over pneumatic tubing. The tubing is usually ¼″ O.D. polyvinylchloride or copper, although sometimes ⅜″ O.D. tubing is used. Catalogs of tube fittings describe fittings to connect ¼″ tubing to tees, elbows, manifolds, and for other tubing sizes as well as to National Pipe Thread (NPT) fittings.† Bundles of tubing are sometimes used to transmit multiple signals from a process to a control room. This is a safe medium for use in environments that may be hazardous (e.g., petrochemical) when using an electric medium. Air pressure is the most convenient power source for operating control valves. The frequency response of a pneumatic control valve actuator cuts off at about 0.3 Hz, whereas a worm-gear driven, electric actuator cuts off at about 0.03 Hz, and it is more expensive than the simple pneumatic actuator. However, the cost of an air compressor system may make the pneumatic actuator more expensive than the electric actuator. Because of slow response, pneumatic signals are not usually transmitted over distances greater than 100 ft.

---

*For more information about radio communication media, see *Radiomodems and Radio Telemetry Eliminate Costly Wireline Data Communications*, published by Ritron, Inc., 505 West Carmel Drive, P.O. Box 1998, Carmel, IN 46032.

†A good example of this kind of catalog is *Fluid Transmission Components,* published by Imperial Clevite, Inc., 6300 W. Howard St., Niles, IL 60143.

Electric Wires

Analog signals (usually 4–20 ma DC) and digital signals are transmitted over electric wires. Each analog signal usually has its own twisted pair of wires to transmit its signal from the sensor to the computer system. The two wires of the twisted pair are usually 18 to 22 gauge copper, insulated from one to the other for at least 100 V, with a shield and an overall insulating jacket that insulates the shield for at least 1000 V to ground. If a cable of many twisted pairs is used, then a single shield with an insulating outer jacket can be provided around the pairs. Each pair in the cable has a different lay or length of twist from all the other pairs to minimize crosstalk between pairs. From a process control point of view, there is no significant time delay or lag in an electrical transmission system, since the speed of light applies to the electrical signal. However, capacitance and inductance in the circuit may slow the signal.

Digital signals are also transmitted over electric wires. For example, a Foxboro I/A transmitter includes a 4–20 ma analog signal on a single twisted pair with the 24 V DC power, as was done in the past. But now digital FSK (Frequency Shift Keying) tones are superimposed on the same twisted pair for bidirectional communication. New digital standards are being prepared for use within a factory. The ISA standard SP50, called Fieldbus, and the HART (Highway Addressable Remote Transducer) protocol are most notable.*

One vendor's standard devices send half duplex digital signals. (Half duplex signals travel in one direction only at any time, as opposed to full duplex travelling in both directions at the same time.) These signals come from up to 32 tank sensors at 1200 baud (bits per second) over a single twisted pair of 18 gauge wire up to 5000 meters long. In this case an FSK modem is used at each sensor with mark/space frequencies of 1200/2200 Hz. A mark signifies a bit or a one when the frequency is shifted to 1200 Hz. A space signifies a zero when shifted to 2200 Hz.†

In order to use a small number of wires for a great many messages between a network of devices, transmission lines have been developed for high-speed transfer of serial digital messages. Coaxial cables as used for cable television are beginning to be used to carry these messages. High speeds mean that the lines must transmit at high frequencies. A broadband system sends many messages simultaneously over a cable using frequency division multiplexing to separate the messages. A baseband system sends one message at a time (at speeds up to 10 Mbaud) and uses time division multiplexing to separate different messages.

Radio Waves

SCADA (Supervisory Control and Data Acquisition) systems may use radio waves to transmit signals from RTUs (Remote Terminal Units) to a central station. These signals are usually digital and are encoded by microprocessor and modem for trans-

---

*The literature for this device is "PSS 2A-1A1 C" in *821 Intelligent Pressure Transmitters,* published by the Foxboro Company, 33 Commercial Street, Foxboro, MA 02035.

†See "Tank level and temperature" in *TGS-11 Tank Transmitter,* published by Datek, Inc., 8522 Davies Road, Edmonton, Alberta, Canada T6E 4Y5.

mission. Usually several sensors are connected to the RTU, and their values are sent as a group to the central station. The RTU may handle 15 8-bit analog inputs and will convert them to digital for serial transmission at 1200 baud over line-of-sight distances at frequencies in the radio bands of 150.8–174 Mhz, 450–475 Mhz, or 928–960 Mhz using half or full duplex.*

Optic Fibers

Optic fibers are used to send signals at very high rates of data transfer over distances up to several miles in the face of extreme magnetic and electric interference. They also offer intrinsic safety where hazards from sparks associated with electric wires exist. The messages are usually digital in format and include signals from many sensors and groups of sensors connected in a network. For example, most distributed control systems, including most programmable controllers, can use optic fibers to interconnect each station over an optic fiber baseband network that transmits messages at up to 10 megabaud (Mbaud). The optic fiber is 100 or 200 microns in diameter, made of glass or plastic, clad with a similar material, strengthened with a tough buffer material (e.g., kevlar), and then protected with a plastic jacket (e.g., polyurethane). Multiple fibers may be cabled together. Special transceivers are required at each end of the fiber, one to convert the optical signal to an electric signal and one to convert the electrical signal back to an optical signal.

4.3 ANALOG FORMAT

An analog signal represents a variable. It has a linear range of variation that represents 0–100% of the possible variation in the variable that it represents. For example, a pneumatic analog signal varying from 3–15 psig may represent the liquid level in a tank that varies from 0–2 m in height. The pneumatic signal can transmit its signal in a ¼″ tube for about 300 ft and maintain the pressure at its correct value, corresponding to the variable that it represents, except for a time lag of about 1 second behind the represented variable. The analog electronic signal of 4–20 ma DC can be sent over a few kilometers of wire with less than a few milliseconds of time lag.

Electronic

The most common form of electronic analog signal uses 4 to 20 ma equivalent to 0–100% of the measured variable. This signal is usually transmitted over wires with less than 200 Ω of resistance. One thousand meters of #22 AWG twisted-pair copper conductors at 25°C have a loop resistance of 108 Ω. Consequently, the voltage drop in 1000 m of this transmission line is 4 V maximum. The supply voltage

*In addition to the literature published by Ritron, Inc., in the footnote on page 87, also see *Honcho Remote Terminal Unit,* published by Repco Radio Canada Ltd., 1060 Salk Road, Unit 6, Pickering, Ontario, Canada L1W 3C5.

is usually from 24–30 V so that little is lost due to transmission losses. Yet it can be made intrinsically safe. Usually the power supply is at the computer location, so that a single twisted pair of conductors carries the power and the signal for one sensor (transmitter). Twisted pairs at worst have a capacitance of 100 picofarads (pF) per meter. For this reason, a 1000-m run would have a capacitance of 0.1 microfarad (μF) and with a 100 Ω resistance would have a time constant of 0.01 millisecond (ms) and could readily pass frequencies up to 10,000 Hz. For most processes 10 Hz is adequate. The separate twisted pair with an individual 4–20 ma signal for each sensor gives very satisfactory results for chemical process measurement and control.

Figure 4.1 is a schematic diagram showing many loop sensors (transmitters), each transmitting 4–20 ma signals from their process measurement locations over shielded, twisted pairs to a central control room. Note that each circuit is grounded at one point only, in the control room, where its shield is also grounded, again at one point only. Also note that it should be possible to lift the ground connection and detect whether any undesired grounds exist. Some of the sensors (transmitters) using thermocouples and RTDs are operated by means of local power. In those cases it is necessary to ensure that the power is isolated with a transformer so that the sensor signal circuit will only be grounded at the control room instrumentation ground. It is very undesirable to have more than one ground on a circuit because of the possibility of circulating currents between the grounds due to lightning or heavy electric machinery drawing high currents. A single ground is desirable to ensure that the circuit insulation is not highly stressed and that the circuit and its shield are at the same potential.

Pneumatic

The standard transmission signal in North America since 1930 has been 3–15 psig (20–100 kPa gauge in Europe). The minimum value or zero value of the signal to be transmitted corresponds to 3 psig, and the maximum value, or full-scale value, corresponds to 15 psig. The signal is usually transmitted in ¼″ O.D. (3⁄16″ I.D.) tubing. One hundred meters of this tubing is a long distance—with a significant drop-off in transmission frequency (breakpoint about 0.2 Hz). For other lengths, see Figure 4.2. In this figure the tubing is assumed to have no fittings restricting the flow. The measured variable is specified to have an error less than 0.5% of span in most cases. The pneumatic amplifiers can usually resolve the 12 psig span to at least 0.1% of span. Existing pneumatic signals are often converted to electric, for computer inputs, using P/I (pressure-to-current) converters that give 4–20 ma corresponding to the 3–15 psig.

4.4 DIGITAL FORMAT

Most new sensors—called *smart transmitters*—incorporate microprocessors. Therefore they can readily use digital signals to communicate with a computer. To remain compatible with existing analog devices, however, they also provide an analog

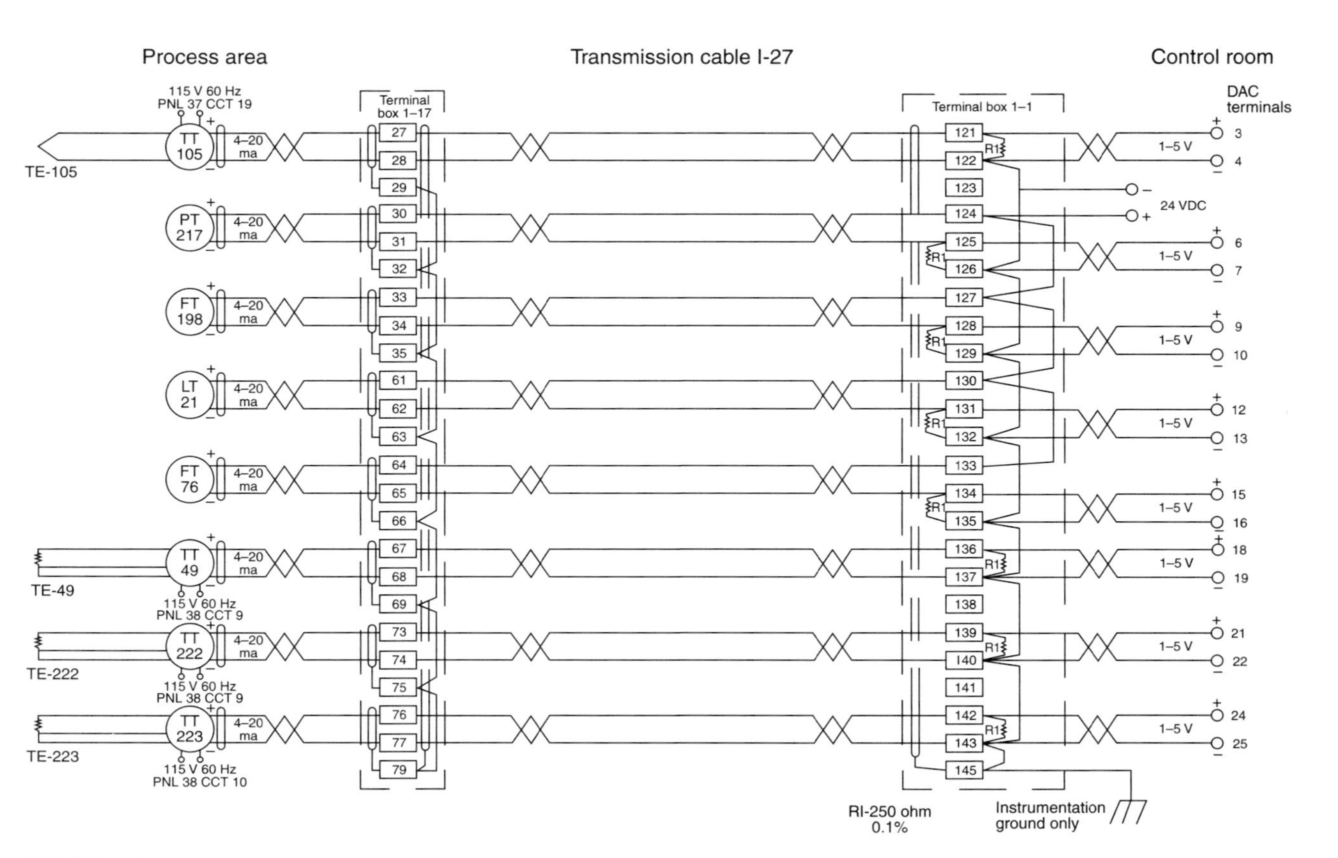

FIGURE 4.1
Analog electric loop schematic diagram

4–20 ma signal. The digital signals are usually the serial type, and they may use electric wires or optic fibers as transmission media. Generally, the digital and analog signals coexist with the power on the same twisted pair of wires. Optic fibers require local power at the transmitter.

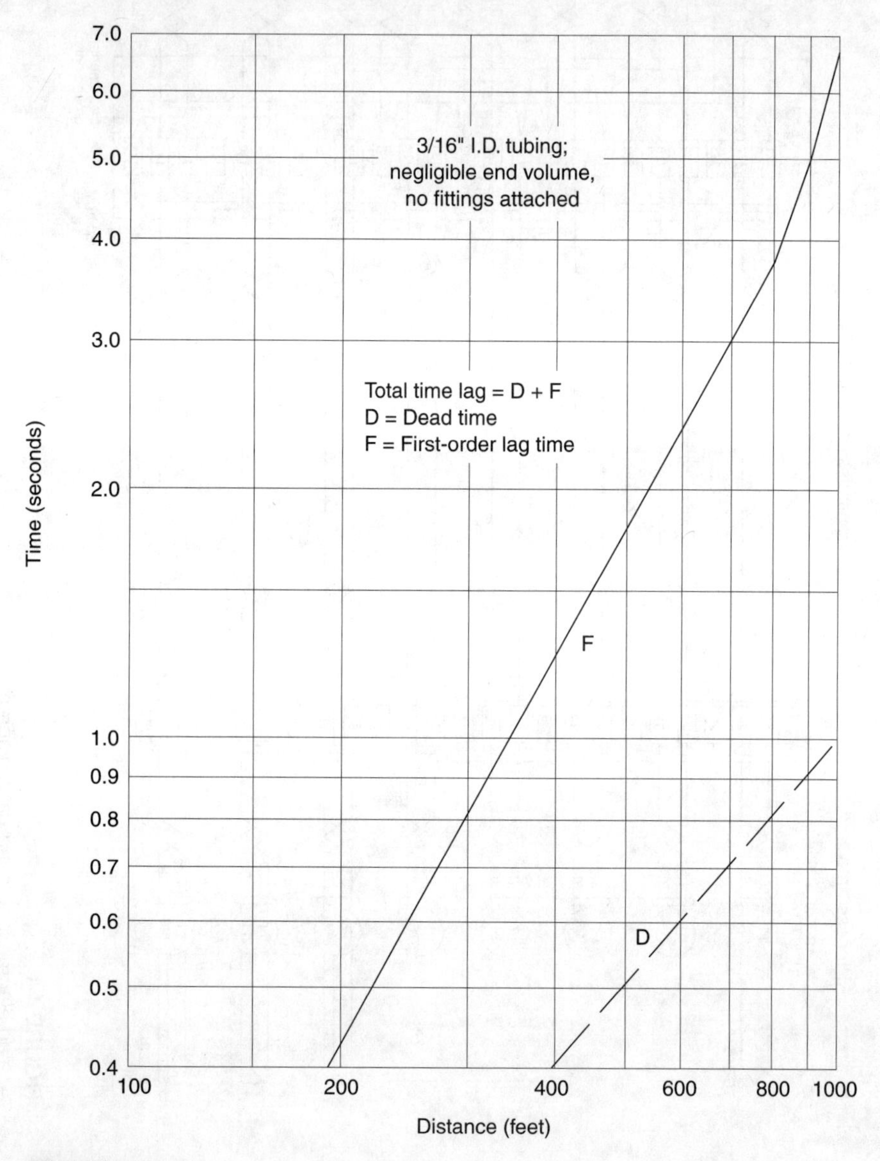

FIGURE 4.2
Pneumatic tubing—transmission lag (Copyright © ISA. Reprinted with permission.)

Parallel

A digital value may be transmitted almost instantaneously from its point of origin to the point of use if it uses parallel transmission. This method of transmission requires a wire for each bit of the digital value plus a return wire. This forms the data bus, but a control bus of wires is also needed. So, for 16 bits of data, at least 17 wires plus about another five wires must run from the point of origin to the point of use. Parallel transmission moves data quickly—but at the expense of many wires.

Serial

Serial transmission uses only one twisted pair of wires between the point of origin and the point of use. To send an 8-bit byte (one character on the screen requires a byte) of data, each bit is sent one after another along the twisted pair. In addition, a couple of extra bits are required to designate the beginning and end of the data. Thus, at 1200 baud (or 1200 bits per second), about 120 bytes are sent each second. All the data and its control information are sent over the one twisted pair of wires. It takes longer to send the information than for parallel transmission, but the initial cost of the equipment is less.

The voltage levels, connector pins, signalling rate, and effective signalling distance for serial communication of digital data over twisted pairs are described in the R/S-232 and R/S-485 Standards of the Electronic Industries Association. The digital data (zeroes and ones) are sent by changing voltage levels from a space to a mark. For example, a space (0) may be a voltage level between +3 and +15 V with respect to signal ground, and a mark (1) may be a voltage level between –3 and –15 V with respect to signal ground. In order to transmit and receive at the same time (full duplex operation), we need two twisted pairs, one wire in each twisted pair going to the signal ground. If only one twisted pair is used, we must wait until we receive a complete transmission before we send our message (half-duplex operation).

In Appendix D, the ASCII (American Standard Code for Information Interchange) characters and digital codes are described. There is a maximum of 256 different characters possible using one byte (8 bits). The 26 lower-case letters (and 26 upper-case letters) of the English alphabet, the standard ten numerals, punctuation characters, special letters from other alphabets, and special control functions make up these 256 characters. It is possible to get by with only the first 128 characters, thus using only 7 bits of the byte that is transmitted. The decimal and hexadecimal value of the byte for each character is shown in Appendix D. By converting the hex value to binary, we can see how the bits appear in each byte that is transmitted. For example, the upper-case letter F has hex value 46, which is 0100 | 0110 in binary. The zeroes would be sent as spaces (+3 to +15 V) and the ones as marks (–3 to –15 V), one after another in time on the single twisted pair.

Each serial port uses a UART (Universal Asynchronous Receive/Transmit) chip to receive and send data over the twisted pair. The serial ports at each end of the twisted pair must interpret the bits in the same way. The receiving port must be sent the information regarding the speed and the bit format of the transmitting port, including the start and stop bits and any parity bit. Otherwise, the message will appear as garbage. The parameters that the UART serial port require are ones that are used for modems. They are usually in the following form:

RRRR,P,D,S,COM?

where RRRR is the decimal value of the baud rate (bits/second) in values starting at 300 and increasing by doubling to 600, 1200, 2400, 4800, 9600, 19800, etc.

D is the number of data bits for each character; either 7 or 8.

P is the parity; either E for even, O for odd, or N for none. For example, if even parity is specified, then the parity bit is included for each character (byte) transmitted to detect transmission errors. If the number of one bits in the data bits is odd, then the parity bit is set to a one so that an even count will result. If the number of one bits in the data bits is even, then the parity bit is set to zero so that, again, an even count will result.

S is the number of stop bits; either 1 or 2.

COM? specifies the serial port of the computer that is being used to transmit or receive the signal; usually either COM1, COM2, COM3, or COM4.

As an example, suppose that an F is being sent over COM2 at 28,800 baud, using 8 data bits, 1 stop bit, and even parity. The parameter specification would be:

28800,E,8,1,COM2

There will be 11 bits for this F character (8 data, 1 parity, 1 start, 1 stop) with timing as follows:

```
serial bit order     | |L| | | | | | |M| | |
serial bit timing     T 0 1 1 0 0 0 1 0 P S
```

where L is the least significant data bit; M is the most significant data bit; T is the start bit (0), P is the parity bit (1), S is the stop bit (1)

The marking (1 bits) is at –5 V values and the spacing (0 bits) is at +5 V values.

If, rather than voltage levels, Frequency Shift Keying (FSK) is used, then its 1200 Hz logical "1" and its 2200 Hz logical "0" have DC, or average values of zero, and do not affect the analog value.

The digital signal allows multi-drop networking so that many devices (measuring sensors and process adjusting devices) may be connected to a single pair of wires. Of course, the many analog signals cannot be allowed to appear on the same pair of wires. Each device has a digital address to which it responds.

4.5 ENVIRONMENT

Process control signals are affected by a number of environmental factors. These are described in detail in the following sections.

Interference

Electrical analog signals can receive significant electrical and magnetic interference along their transmission lines if they are not made up as twisted pairs. The same is true if they are not shielded, or if they are not grounded at one central location only.

Cables used for telephone applications are also effective for transmitting 4–20 ma DC signals from process sensors to a computer system. The telephone cables include multiple (usually 50 or 60) twisted pairs of 18 to 22 gauge wire, insulated for at least 100 V, with each pair twisted at a different lay, and all inside a common aluminum shield with an overall insulating jacket.

A major form of interference is due to circulating currents in the twisted-pair circuits or in the overall shield (see Figure 4.1). If a twisted pair is grounded in the control room and also at its sensor, then circulating currents can occur through the wires and ground, because the two grounds could be at different voltages due to ground currents from heavy electrical machines or from lightning. Each circuit and the shield must be isolated from ground everywhere except at one point. At the central computer location—usually the control room—a separate instrumentation ground electrode is run into the earth. No power system circuits should use this ground. Only the instrumentation cable shields and instrumentation circuits should use it. All cable shields must be kept insulated everywhere except in the central location. Each circuit must have the shield over its twisted pair connected back to the central computer ground terminal only. Its shield should be connected via all the shields that are over its own twisted pair all the way from the sensor via closer shields right up to the central instrumentation ground terminal. The existence of more than one ground may be checked for by temporarily lifting the ground connection to the central instrumentation electrode. If the resistance to ground is much less than a few megohms, then an undesired ground may have been connected inadvertently.

The twisted pair from each sensor must be well insulated from ground, except at the central instrumentation ground. The sensor itself must have its circuits isolated from ground. If the sensor is operated from its own local power source, then it must be isolated from the ground of that power source by means of an isolating transformer. The circuit for each sensor must have its twisted pair grounded at only one location, the central instrumentation ground. This single ground ensures that no circulating currents appear and that the circuit is not floating at a high voltage. It also ensures that the shield and the circuits inside the shield are relatively close in voltage. If these features are not provided, then possible damage to the devices and to the cable insulation is likely, especially as a result of a lightning strike near the factory.

In order to minimize interference from power cable magnetic fields, the sensor signal cables must be run in trays separated from any power cable runs by at least a meter.

These techniques will overcome most analog circuit interference within large process plants. Signals of 4–20 ma DC in runs up to at least 2000 meters function satisfactorily in such environments.

Safety

Hazardous vapors or dust may explode if ignited by an electric spark. Special precautions are required for electrical sensors and their transmission circuits that operate in such hazardous environments. Electrical sensors are often designed to meet intrinsically safe operating requirements that have been established by UL (Underwriters Laboratories), CSA (Canadian Standards Association), and FM (Factory Mutual). These devices may operate inside the hazardous area, but must operate on voltages of less than 30 V, must not store electric energy that could cause a dangerous high-energy electric spark (very weak sparks will not cause explosions), and their circuits must be protected from receiving excessive electrical energy from their power supply by means of electrical safety barriers that are located outside the hazardous area. In order to be approved intrinsically safe, such a circuit must include a sensor or transducer designated intrinsically safe, must operate from a power source that produces less than 30 V, and must include safety barriers in a nonhazardous area. Safety barriers include fuses, diodes, and resistors to ensure that minimal current—just enough for normal operation—flows to the sensor or transducer in the hazardous area.*

Intrinsically safe circuits (see Example 4.1) are somewhat less expensive to construct than the alternative explosion-proof circuits. If an area is designated hazardous, as classified in the appropriate electrical code, then all electric circuits in that area must be made intrinsically safe or be placed inside explosion-proof fittings. Any explosion due to sparks from the electric circuits will only exist inside the explosion-proof fittings, since the high-energy sparks can only occur inside the fittings, and the explosion-proof fittings are designed to contain the explosive gases without rupturing.

EXAMPLE 4.1 **Intrinsically Safe Level Measurement of Gasoline Tank Level**

Measurement of the level of gasoline in an outdoor, above-ground, fixed-roof tank requires an intrinsically safe electrical instrument. The Canadian Electrical Code 20-302(4)(b) and the U.S. National Electrical Code articles 504 and 515 specify for class I (flammable gases or vapors), Division 2 (in hazardous concentrations that do not normally exist but could exist as the result of an accident), Group D (gases and vapors from petroleum liquids or gases), the hazardous area extends for 3 m (or 10 ft) outside the shell, ends, and roof of the containing vessel.

---

*For additional information and for classification of hazardous locations, see "Recommended Practice RP 12.6," *Installation of Intrinsically Safe Instrument Systems in Class I Hazardous Locations,* published by the Instrument Society of America.

Equipment specification for Example 4.1:
An intrinsically safe, flanged, electronic level (differential pressure type) transmitter shall be mounted at the bottom of the tank to measure the pressure and the level of gasoline in the tank. The 4–20 ma DC signal from the transmitter shall be connected via a shielded twisted pair of #22 AWG stranded wire to two intrinsic safety barriers (one for each conductor) located outside the hazardous area. The signals shall pass through the barriers to the refinery distributed control system for presentation and recording of the tank level. A power supply of less than 30 V DC, located in the non-hazardous area, shall be used to excite this circuit.

Pneumatic transmission systems do not cause sparks and are therefore considered inherently safe. Optical systems are also considered inherently safe.

Junction Boxes

Junction boxes are used for each type of transmission media to collect signal lines from each sensor and cable them together to run them back to the central data acquisition area. For electric cables, each shield requires a separate terminal so that insulated shielded connections are made, and so that the shield can be disconnected when searching for undesired and accidental shield grounds. Consideration should be made for ensuring that dust and water do not cause problems for the connections while the box is closed as well as when service personnel have the box open. For thermocouples running back from the field to the junction box, care must be taken so that reference junctions are established only where they are required.

Special care is required when optic fibers are terminated. Splicing these cables requires special tools and splicing kit.

For pneumatic tubing there are union fittings that pass through a junction box so that individual tubes may interconnect with a bundle of tubes cabled together. In this way, the pneumatic field devices are efficiently connected to the control room. It is not desirable to have these runs longer than 100 m because of the delay in pneumatic signals.

Air Supply

Instrument air is compressed and conditioned to power pneumatic sensors, controllers, and actuators. To ensure accurate and reliable operation of these devices, the air must be clean, dry, and free of contaminants, and it must remain at a constant pressure. The preferred humidity has a line pressure dew point at least 10°C below the minimum ambient temperature and not above 2°C. The oil content of the air should be less than 1 part per million (ppm), and it should be filtered so that any solid particles will be smaller than 3 micrometers (μm) in diameter. The air supply pressure at each device must be regulated to 20 psig. Often several closely grouped devices use one filter regulator to drop the air pressure to 20 psig from the distribution line pressure of the factory, usually between 60 and 100 psig. If the devices are spread apart by more than 2 m, it is desirable for each field device to be equipped with its own air pressure filter regulator.

Electric Power Supply

Reliable, constant-voltage electric power is required for operation of a data acquisition system. The battery operated UPS (uninterruptible power supply) is becoming desirable in many applications. In this case the battery is continuously charged by the local electric power utility, and if the power fails, then a solid state switch automatically removes the power utility and the battery supplies the control system until the power returns. Batteries are sized for a certain number of minutes of operation without the charging power from the utility. The time should be long enough to allow safe shutdown of the process. For some special processes, such as nuclear power stations, fault-tolerant power sources are required, and these may include two or three sources, such as the UPS and a diesel generator.

Solar Power

In remote locations, such as a metering point on a pipeline hundreds of miles long, a signalling location on a railway in a mountain region, or a meteorological station gathering weather data, equipment for the acquisition of data can be established. In such locations, reliable electric power is required as inexpensively as possible. Running electric power lines on poles is expensive and probably would not be implemented. However, batteries charged by solar cells may make such a project feasible. Standard units for roof or pole mounting include solar cells and batteries that provide up to 125 ma at 12 V continuously.* With fully charged batteries (28 ampere hours) the load can draw 125 ma for at least 224 hours without solar recharging. The 9 W (at peak sunlight) solar cells will charge the discharged batteries at 0.75 A (9 W/12 V) in 37.3 hours at peak sunlight and no useful load. Thus 24 hours × 0.125 A/0.75 A = 4 hours at peak sunlight—or 8 hours at 50% sunlight—per day will keep the batteries charged continuously and have a reserve, when charged, of more than nine days.

4.6 NETWORKS

Before DCS (distributed control systems) were common, analog signals were standard. Networks did not exist. When DCS began to appear, the users saw a proliferation of many networks, each with its own method of communication between stations.

---

*Descriptions of solar power devices are in *Photovoltaic Products,* published by Solarex Corporation, 1335 Piccard Drive, Rockville, MD 20850; and *ST-10 Photovoltaic Power System,* published by Integrated Power Corp., 7524 Standish Place, Rockville, MD 20855.

The users did not want to be tied totally to one vendor and its method of communication, so they formed groups to standardize the way devices communicate with one another. The vendors also saw advantages in standardizing, so they joined these groups, and strenuous efforts have been made to standardize. Many organizations around the world are now trying to agree on a minimum number of standard ways of communicating. A major network that already exists is the telephone network. Telephone engineers have solved—or at least described—many of the problems that data acquisition and control systems now face regarding the method of communicating on a network.

Telephone System

The telephone system is sometimes used as part of a data acquisition system. It can be very convenient to establish several sensors that are distant from the central computer station and transmit their readings frequently over the telephone system for presentation to the central computer. For example, a few years ago an Ontario housing research group wanted data on the energy use and thermal insulation performance of a group of houses of a new design. They wanted temperatures inside and outside the houses, electric power usage, and furnace gas or oil usage. They wanted this information reported daily every 15 minutes for six different houses then averaged and reported monthly. Looking back now, an ideal way to do this would use a personal computer in each house with an analog input interface to a sensor for each variable. Each computer would include a modem connected to its own phone line. A central computer at the research office would call each house computer once a day to receive all the day's data collected by it for the previous 24 hours. This way the research office would have an up-to-date record of the house operation and performance every day. Other applications include data from weather stations, security data from unattended buildings, remote operation of unattended hydroelectric power stations, and the like. The weather station on top of a mountain might require a solar power supply and a radio communication medium to a telephone system modem.

Radio Networks

The Ritron RadioModem allows PLC stations to communicate as point-to-point links or polled multi-station networks.* It acts as a telephone modem, so that no new way of interfacing the PLC station to another PLC station is required. The connector is the same serial connector required to interface to a modem as is used on the telephone system. It can operate over a range of 10–20 km in full or half duplex configuration at 1200 baud as an alternative to wire or fiber optics. It may be an ideal way to operate a small process, such as remote operation of a plant overhead crane from the manufacturing floor. However, the 1200 baud rate is much too slow for any more than a dozen variables.

---

*See the footnote on page 87 for information on how to obtain Ritron, Inc., literature on this product.

Local Area Networks

The acronym LAN (local area network) is a computer term to describe multiple computer stations that can communicate with one another. Local means in one building or one plant, so this would apply to a process with many control loops. Many LANs have just grown. They have used the standards that exist and have not tried to have suppliers conform to any standard. These networks have made do with simple twisted pairs connecting PCs, PLCs, and instruments as a group.

MAP/TOP

Two major attempts to bring standardized LANs to the factory and to the office are MAP and TOP. MAP stands for Manufacturing Automation Protocol, which was initiated by General Motors in the early 1980s to provide standardization of factory network communication system hardware and software. This standardization is necessary to allow intercommunication between various manufacturers' devices. TOP stands for Technical and Office Protocol initiated by Boeing Computer Services in the early 1980s for similar standardization in office and engineering environments.

Vendors want to sell their equipment to users in many countries. Users want to be able to buy from a variety of vendors from various countries. By standardizing the kinds of signals and power supplies that the equipment requires, everyone benefits. The International Standards Organization (ISO) developed a reference communications network called the Open Systems Interconnect (OSI), which was selected by MAP as its model. The OSI is said to have seven layers, whereby messages originating in an application program of a vendor's device are translated by the various layers to signals for communication via a transmission line with other devices. The various layers, not all of which are essential for every device, are shown in Table 4.1.

Many groups have become involved with MAP, and MAP has attached itself to many other groups. In order to achieve enough agreement to make this concept work, a great deal of time and paper has been devoted to the subject. Eventually the most acceptable and probably the most productive method for intercommunication will result. As the Process Industries Initiative Working Group of the MAP/TOP Users Group of the SME Secretariat observed:

> The long term goal for MAP is to provide a communications network that allows applications to share information through standard protocols at all applicable levels of plant operations. The final analysis of MAP will be based on its ability to meet performance requirements of the process industry in a cost effective manner while supporting real-time applications, providing reliable service, and providing a clean migration path from already existing process device networks.*

The physical link of a LAN is the serial communication line, which uses one of three kinds of frequency bands: baseband, carrierband, or broadband. A baseband signal is the original single channel of information without modulation having each bit represented as a particular signal level. A carrierband signal is the original single bi-

*MAP Process Industries Initiative Working Group. 1987. *MAP in the Process Industry.* Dearborn, MI: MAP/TOP Users Group of the SME Secretariat.

TABLE 4.1
Open system interconnect layers

Number	Layer Name User Program	Function User Application	MAP Specification
7	Application service	Directly services application program	ACSE, FTAM, MMFS/EIA 1393A
6	Presentation	Transforms data to/ from standard format	Kernel, ASN.1
5	Session	Synchronizes and manages dialogues	ISO session full duplex
4	Transport	Monitors reliable transport from end node to end node	ISO transport class 4
3	Network	Packet routing for data transfer between nodes	ISO connectionless network service
2	Data link	Error detection for messages moved between nodes	IEEE 802.2 link level control class 1
1	Physical	Encodes and transfers bits between nodes	IEEE 802.4 token access on broadband media
0	Physical link (transmission line)		

Source: MAP Process Industries Initiative (MPII) Working Group. 1987. *MAP in the Process Industry.* MAP/TOP Users Group of the SME Secretariat, 1 SME Drive, P.O. Box 930, Dearborn, MI 48121, (313) 271–1500.

directional channel of information modulating a carrier frequency. A broadband signal carries several channels of information, such as voice, data, and video, each channel modulating a different carrier frequency on the line. The broadband and carrierband signals represent a bit by modulating the carrier amplitude or frequency with a tone or by phase shifting the carrier. The modulated signals are usually less susceptible to industrial noise than the baseband signal. MAP has not laid down which method of physical link must be used.

In one factory there could be several networks—some proprietary, such as Ethernet (developed by Xerox)—plus the open, standard backbone MAP broadband network, and other MAP carrierband subnetworks. Communications relays are required to provide internetwork communications. There are three kinds of communications relays: bridges, gateways, and routers.

A bridge relay interconnects two similar networks, for example, two MAP carrierband subnetworks. A gateway relay interconnects two networks that are totally different, such as an Ethernet and a MAP network. A router relay connects multiple networks together.

It is fairly simple for digital equipment to provide for one master station on a network and multiple slave stations on that network. Each slave station only transmits data when it is requested—or *polled*—to do so by the master station. Major problems appear when multiple peer stations exist on one network if any station is allowed to transmit to any other station or group of stations at any random time. This is the case with digital networks. Two methods, token passing and collision detection, have been devised to solve the problem of more than one station transmitting on the network at the same time.

The *token passing method* makes each station the master for a short time. When a station is the master, it is said to hold the token. That station can then transmit to whomever it wants. If it no longer has anything to transmit, it must pass the token to the next station. The token thus gets passed around and around and messages are transmitted with a moderate amount of overhead, but a known amount.

Another method is called the *collision detection method.* It allows all the stations to transmit at any time. All stations listen to the network at all times and will only transmit if the network is idle. However, there is the possibility that more than one station will commence transmission at exactly the same time. The resulting collision is detected by the transmitting stations, and then they listen to the network and after a random time recommence transmission if the network is idle. The overhead associated with a low utilization of this network is very low. But as the network becomes more and more utilized, there is a greater likelihood of collisions and a greater amount of overhead associated with each message.

Network problems are complex and a final few types of networks have not yet been agreed upon. This major feature of process control will be a hot topic for a long time.

Fieldbus

The Instrument Society of America is working with many international standards organizations to establish a new specification, SP50, which will govern the media and format of digital data transmission for process measurement and control. The draft for this specification calls it *Fieldbus.*

A demonstration physical layer for a high-speed Fieldbus, called ISIbus by British Petroleum,* uses a twisted pair that includes AC power for many field sensing devices and actuating devices along with their serial, digital signals. The AC power and signals are inductively (transformer) coupled to each device. The twisted pair is not broken at the device; instead a single-turn transformer, similar to a clip-on ammeter, couples the device to the twisted pair. It is operating for up to 20 field devices

---

*A description of this device is in *ISIbus* by Bjørn Raad, published by British Petroleum and Senter for Industriforskning, P.O. Box 124, Blindern, 0314 Oslo 3, Norway.

providing 100 mW per device at 1 V with a frequency of 14 kHz and transmitting digital signals at 1 Mbaud over 500 m of twisted coaxial cable.

Several such buses exist. Some are open, while others are proprietary. As with MAP, these networks and buses are still in a rapidly evolving state. Valuable features will eventually come into being, and in a few years these will become standard. Until then uncertainty will exist on every path that is selected.

PROBLEMS AND LAB ASSIGNMENTS

4.1 Every day a check is made on the instrumentation ground of Figure 4.1 on p. 91 by lifting the ground from terminal 145 and ensuring that the resistance between the ground and terminal 145 is in excess of 1 MΩ. If you found a resistance of 65 Ω, what procedure would you follow to locate the offending ground? As a precondition, suppose that the ground is on the shield of the 4–20 ma twisted pair at the TT-105 sensor, but you do not know this to start with. As a separate precondition, suppose the undesired ground includes the shield and negative terminal of PT-217.

4.2 From Figure 4.2 on p. 92, find the dead time plus five time constants for ¼″ O.D. (3⁄16″ I.D.) pneumatic tubing for lengths of 100, 200, 300, 400, and 500 ft.

4.3 Draw a diagram similar to Figure 4.1 for Example 4.1 on p. 96 of an intrinsically safe installation. Show a safety barrier as a small three-terminal device with an incoming current, outgoing current, and ground. Use dotted lines to delineate the hazardous area from the nonhazardous area.

4.4 Draw a schematic diagram similar to Figure 4.1 for the example of an intrinsically safe installation described in Problem 4.10. Show a safety barrier as a small three-terminal device with an incoming current, outgoing current, and ground. Use dotted lines to delineate the hazardous area from the nonhazardous area. Show typical ISA instrument tag designations and typical terminal numbers. Show a separate schematic diagram for the 550 V motor with circuit breaker, motor starter, and push buttons.

4.5 A campsite operator wants to light his parking lot near a remote lake with sodium vapor lamps mounted on poles that include batteries and solar cells, a darkness sensor, and a time-on timer. If he uses a 12 V, 35 W lamp on each pole with a 12 V, 48 W at-peak sunlight solar cell module with storage batteries storing 250 A hours, then how long can he set his timer on each set of lights to run each night without running down its batteries, if the solar cells receive an average of eight hours at 50% of peak sunlight each day? How much reserve time does he have in a set of fully charged batteries?

4.6 Fifteen sensors are connected as a simple network on a single twisted pair to a computer, and each sensor has a digital address used by the computer. Each digital request message made by the computer for a measurement value from a sensor has 4 bytes, and each response message by the sensor sent back to the computer has 4 bytes. How frequently can the data be obtained from any sensor, if the data are being sent serially at 9600 baud, and if all 15 sensors, one after another, are receiving request messages and are sending response messages?

4.7 Pneumatic Tubing Transmission Time Constant

Objective: To measure the time constant and delay required to pressurize a long length of ¼″ plastic tubing.

Procedure: Prepare a long (over 150 m) length of pneumatic ¼″ plastic tubing on a reel with a pressure gauge (0–15 psig) connected to each end of the tubing. Estimate the length of tubing to about 10% accuracy. Place a quick-connect connector on one end. Draw a diagram of the tubing connected to a pressure regulating valve by means of the quick-connect connector.

With the tubing not connected to the pressure regulating valve, set the pressure regulating valve to 10 psig. Start timing at the moment that the pressure regulating valve is connected to the tubing. Record the time that the pressure at the far end of the tubing reaches each integer of psig, while it is increasing from 0–10 psig. Repeat the timing while decreasing from 10–0 psig after the supply end of the tubing is disconnected from the regulator.

Conclusions: Plot the pressure (vertical axis) versus time (horizontal axis). Estimate the time constant (time to change 63.2% of the overall change) for transmission along this length of tubing. How does the time constant compare with Figure 4.2?

4.8 Pneumatic Tube Fittings Selection

Objective: To select pneumatic tube fittings from a catalog.

Procedure: Obtain a catalog of tube and pipe fittings.* A small panel of pneumatic instruments is required to be tubed, and you are requested to draw the tubing schematic and select the fittings for this panel. The front of the panel includes a pneumatic indicating flow controller, a separate three-pen pneumatic recorder, and a knob that adusts a 20 psig air supply filter regulator with an air supply pressure indicator. A square root extractor for the incoming flow signal is located behind the panel. The panel requires a pneumatic header supplied from the filter regulator with air supply lines for the square root extractor, the flow controller, and the pneumatic servo-type recorder. The 60 psig incoming air supply enters the panel through a ½″ pipe. The incoming and outgoing pneumatic measurement—tank outflow (orifice differential pressure), tank pressure, and tank level—and control (flow control valve) lines arrive at the panel in a bundle of ¼″ pvc tubes and pass into the panel via tube union fittings. Examine examples of these devices in manuals or in your shop to see typical pneumatic connections. Draw a pneumatic schematic diagram interconnecting the devices correctly, and mark each tube fitting on it that you think is needed. List all the tube fittings with their catalog descriptions.

---

*A good example of such a catalog is *Swagelok Fittings, Nupro and Whitey Valves,* published by Swagelok Company, 31400 Aurora Road, Solon, OH 44139, http://www.swagelok.com

Conclusions: How long did it take you to analyze the problem, draw the diagram, and select the fittings? How long would it take you now to do a similar job for a different panel? What additional service connections are required for the recorder in this panel? Submit your list and diagram with your answers to these questions in your report.

4.9 Digital Serial Transmission RS-232C at Various Baud Rates

Objective: To measure the actual transmission speed of a file sent over a RS-232C serial link at various baud rates.

Procedure: Use a null modem cable to connect two computers through their serial ports and a communication program such as PROCOMM or PC-TALK at 300 baud, 1200 baud, 4800 baud, and 9600 baud to transfer a file (e.g., LICENSE.DOC in PROCOMM) from one PC to the other PC. Find the time to receive the file at each nominal speed and check for any transmission halt, XOFF. More than likely you will be using a serial RS-232C port so that the distance between computers should be less than 50 ft (16 m). From the DOS directory, obtain the number of bytes in the file and assume the number of bits transmitted is ten times this number of bytes.

Conclusions: In a table compare the actual average rates (bits per second) of transmission of the file to the nominal rates.

4.10 Classification of a Hazardous Area

Objective: To establish the boundary of a hazardous area.

Procedure: Your company is performing design work for a gasoline storage installation. Review sections 20-300 to 20-400 of the Canadian Electrical Code or Articles 504 and 515 of the U.S. National Electrical Code. The above-ground outdoor tank is located inside a dike. The tank is 5 m high and 10 m in diameter. It has a fixed roof. Its small unventilated service building (3 m by 3 m) is located just outside the dike near the tank, and it houses:

- Two intrinsically safe, 4–20 ma signal, and pneumatically actuated control valves.
- A pump with a 3 HP, 550 V motor.
- Six 4–20 ma intrinsically safe transmitter instruments.

Draw a layout of this tank and its service building. Draw dotted lines to indicate the hazardous and nonhazardous areas of the installation. Show dimensions. Show the classification that you believe each hazardous area should have, for example, Class I, Division 2, or Class I, Division 1.

Conclusions: Where would you locate safety barriers for the intrinsically safe devices? What wiring equipment would you use to connect the 550 V power to the motor? What wiring equipment would you use to connect the safety barriers to the intrinsically safe devices? How would you ground the equipment in the hazardous and nonhazardous areas?

REFERENCES

Canadian Standards Association. 1986. *1986 Canadian Electrical Code Part I.* Rexdale, Ontario: Canadian Standards Association.

Datek Industries Ltd. 1986. *TGS-11 Tank Transmitter.* Edmonton, Alberta: Datek Industries Ltd.

Foxboro Co. 1988. "PSS 2A-1A1 C," *821 Intelligent Pressure Transmitters.* Foxboro, MA: Foxboro Co.

Instrument Society of America. n.d. "Recommended Practice RP 12.6," *Installation of Intrinsically Safe Instrument Systems in Class I Hazardous Locations.* n.p.: Instrument Society of America.

Integrated Power Corp. 1986. *ST-10 Photovoltaic Power System.* Rockville, MD: Integrated Power Corp.

MAP Process Industries Initiative (MPII) Working Group. 1987. *MAP in the Process Industry.* Dearborn, MI: MAP/TOP Users Group of the SME Secretariat.

Murrill, P. W. 1981. *Fundamentals of Process Control Theory.* Triangle Park, NC: Instrument Society of America, Research.

Raad, B. 1988. *ISIbus.* Oslo: British Petroleum and Senter for Industriforskning.

Repco, Inc. 1986. *Honcho Remote Terminal Unit.* Pickering, Ontario: Repco Radio Canada, Ltd.

Ritron, Inc. n.d. *RadioModems and Radio Telemetry Eliminate Costly Wireline Data Communications.* Carmel, IN: Ritron, Inc.

Solarex Corp. 1986. *Photovoltaic Products.* Rockville, MD: Solarex Corp.

Swagelok Co. n.d. *Swagelok Fittings, Nupro and Whitey Catalog.* Solon, OH: Swagelok Co.

5

PROCESS ADJUSTMENT

OBJECTIVES

When you have completed this chapter you will be able to:

- Define and list the purposes for liquid flow adjustment
- Describe the application of a variable speed pump to liquid flow adjustment
- Describe the application of a control valve to liquid flow adjustment
- Size a control valve for liquid flow adjustment
- Select a control valve for liquid flow adjustment
- Define and list the purposes for gas flow adjustment
- Describe the application of a control valve to gas flow adjustment
- Size a control valve for gas flow adjustment
- Describe the application of a variable speed fan to gas flow adjustment

5.1 LIQUID FLOW ADJUSTMENT

The *adjustment of liquid flow rate* is defined as the ability to vary smoothly the rate of transfer of a liquid in a pipe over a range from a minimum to a maximum. The purposes of adjusting the flow rate of a liquid or a slurry are

- To set the rate of raw material flowing into a process, such as crude oil to a distillation process.
- To set the rate of energy entering a process, such as fuel oil to a cement kiln.
- To set the rate of withdrawal of energy from a process, such as the flow rate of brine to an ice-making process.
- To set the rate of withdrawal of material from a process, such as the rate of pulp blowing from a wood-pulp digester.

Variable Speed Pump Flow Adjustment

Figure 5.1 shows a typical flow process where flow is adjusted by varying pump speed. Liquid from Tank A is sucked into a pump, and then the pump pressure forces the liquid to flow through pipes and a flow measuring orifice into Tank B. In order to understand the liquid flow process, we must relate the fluid pressure rise in the pump and the pressure drops in the pipes and orifice to the flow rate. The pressure drops are like voltage drops in an electrical circuit. But in an electrical circuit the voltage drop increases linearly with increasing current. In pipes the pressure drop increases with the square of the increasing flow. Assume the tanks are open to the atmosphere and that the lowest elevation of the piping is at the exit of Tank A. The pressure, P_a, at this point is due to the head, L_a, of Tank A. There will be a slight pressure drop in the pipe from Tank A to the pump. The pressure rise due to the pump at full speed S can be found from the pump curve S (see Figure 5.2), if the flow rate is known or assumed. The pump pressure rise, or pump head, is $P_{po} - P_{pi}$, that is, pressure at pump outlet minus the pressure at pump inlet. This pressure rise can be found from the pump curve shown in Figure 5.2. In Figure 5.1 the value of 290 kPa at a flow of 83 lpm was taken from Figure 5.2 with the pump running at rated speed, S. At this flow the pump pressure rise just equals the pressure drops in the pipes and orifice plus the tank head, P_a, in Tank A minus the tank head, P_b, in Tank B.

Figure 5.2 shows that if the pump runs at a speed less than the rated speed, then the pump head is reduced. The pump head is proportional to the square of the pump

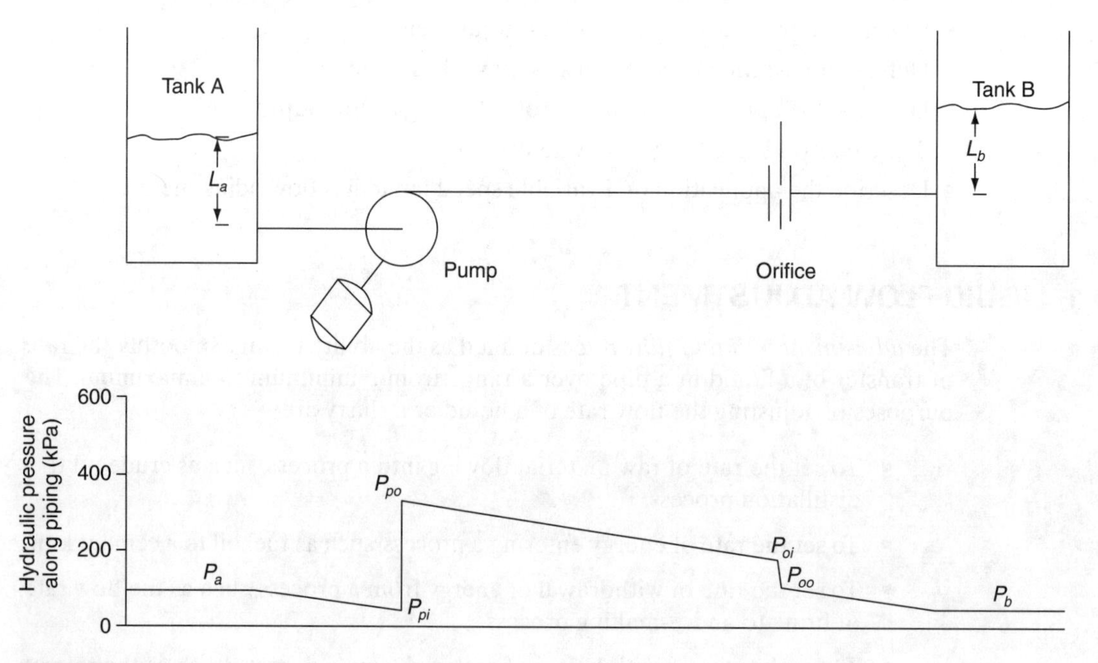

FIGURE 5.1
Fluid pressure drop along piping

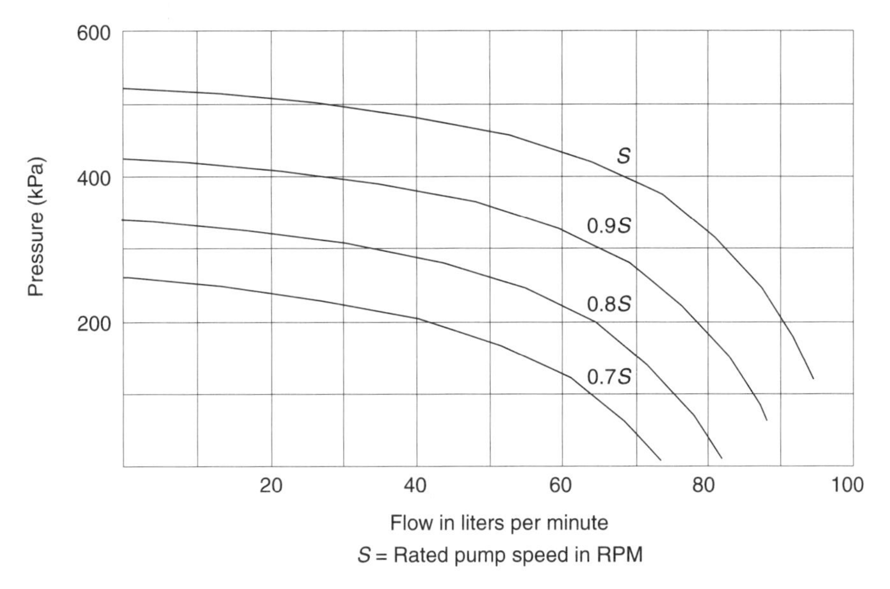

FIGURE 5.2
Pump head vs. flow

speed, so the head curves come closer together as the speed is reduced in equal amounts. At higher flow rates the frictional losses in the pump alter this relationship somewhat.

The practical ranges of flow rates in the piping almost always have a Reynold's number greater than 10,000 and are in the turbulent range, causing frictional pressure drops that are proportional to the square of the flow rate. (This is the function used for curve *F* in Figure 5.4 on page 111). Also, the longer the pipe the greater the pressure drop; in fact, it is directly proportional to the length. Actual pressure drop in a pipe may be calculated as described in Appendix E.*

With a variable speed pump, it is possible to control liquid flow so that no control valve is required. The need for tight shutoff when the system is shut down, however, may demand a shutoff valve, because the pump may windmill or, if you prefer, waterwheel due to fluid pressure when the motor is shut off.

Control Valve Flow Adjustment

Figure 5.3 shows a typical flow process where flow is adjusted by means of a control valve. Liquid from Tank A is sucked into a fixed-speed pump, then the pump pressure forces the liquid through pipes, a control valve, and a flow measuring orifice, and then into Tank B. As before, the pump produces a pressure rise and this pressure rise plus the head in Tank A is just equal to the pressure drops in the pipes, control valve, and orifice, plus the head in Tank B.

*Similar data is provided by Crane Company (1957) and Fisher Controls Company (1977). See the chapter references.

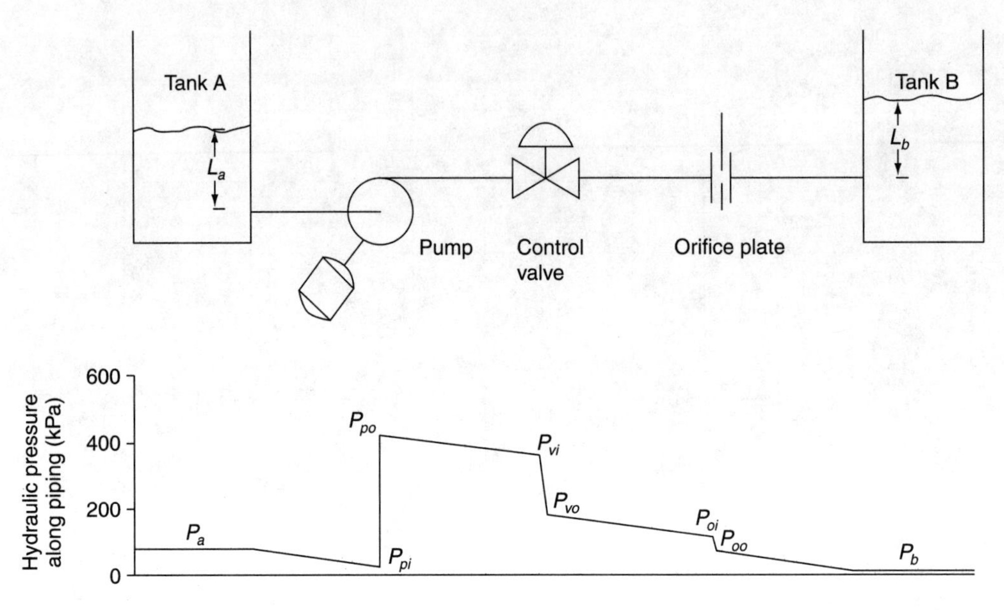

FIGURE 5.3
Fluid pressure drop along piping

Figure 5.4 shows the pressure drop that the control valve in Figure 5.3 wastes to achieve a particular flow rate, *Z*. The inlet pressure to the pump is L_a, due to the level or head of Tank A. The pump pressure rise curve, *S*, from Figure 5.2 has L_a added to it to form the supply pressure curve *C* in Figure 5.4. The outlet pressure of the piping is L_b, due to the head of Tank B. The friction pressure loss in the piping and process has L_b added to it to form the demand pressure curve *F*. To achieve any flow shown on the X axis of Figure 5.4, the control valve must be able to absorb the pressure drop between curve *C* and curve *F* at that flow rate by adjusting its opening to the appropriate value. As shown on Figure 5.4, it can only provide the adjustment between the minimum flow, *N*, and maximum flow, *X*.

Capacity of a Liquid Control Valve

For liquids that do not involve flashing, cavitation, or high viscosity, the ISA equation 5.1 (simplified) for volumetric flow rate relates the valve flow capacity factor, C_v (same numerical value for English and for SI units), to *Q*, the volumetric flow rate through the valve; *G*, the specific gravity of the liquid; ΔP, the valve pressure drop; and K_1, a constant.

$$C_v = K_1 \times Q \times \sqrt{\frac{G}{\Delta P}} \tag{5.1*}$$

*Equation 5.1 and subsequent numbered equations in this chapter are reprinted from "ISA, Standard SP75.01, Control Valve Sizing Equations, 1977." Reprinted with permission. All rights reserved.

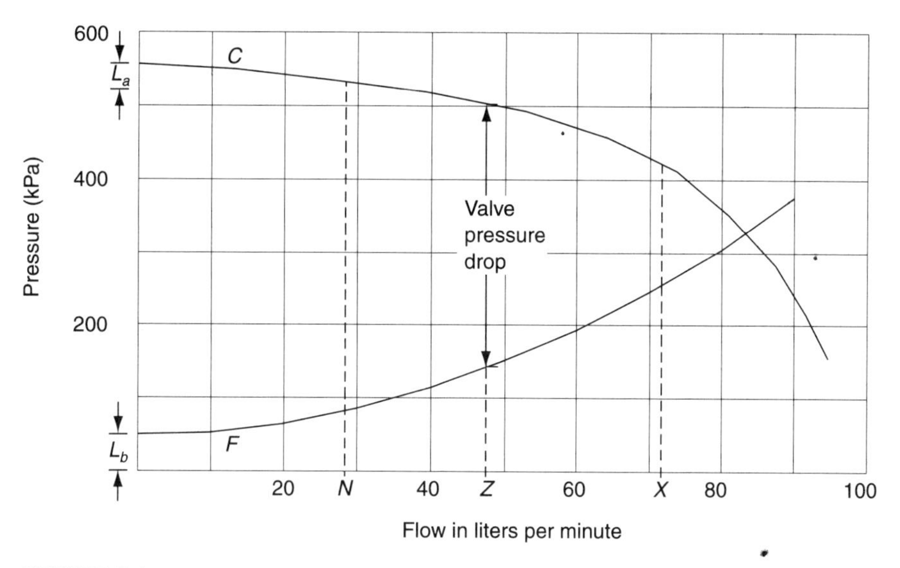

FIGURE 5.4
Control valve pressure drop to achieve a flow rate

where:

English Units	*SI Units*
$K_1 = 1$	$K_1 = 11.56 = 1 / 0.0865$
Q in GPM	Q in m^3/hour
$G = 1.0$ for water	$G = 1.0$ for water
ΔP in psi	ΔP in kPa

For the mass flow rate, *W*, through the valve, the ISA equation 5.2 (simplified) is

$$C_v = \frac{W}{K_2 \times \sqrt{G \times \Delta P}} \tag{5.2}$$

where:

English Units	*SI Units*
$K_2 = 500$	$K_2 = 86.5$
W in pounds/hour	W in kilograms/hour
$G = 1.0$ for water	$G = 1.0$ for water
ΔP in psi	ΔP in kPa

Usually C_v is considered to be the capacity of the valve when it is wide open. However, a smaller value can also be used in the same equation as the reduced capacity when the valve is partially open. When it is necessary to differentiate in this book, C_{vf} will designate fully open C_v, and C_{vp} will designate partially open C_v.

EXAMPLE 5.1 Sizing a control valve for water in 2.54 cm pipe can be accomplished by using Figures 5.3 and 5.4. First, the C_v equation (5.1) is rearranged to find *Q:*

$$Q = \frac{C_v \times \sqrt{\frac{\Delta P}{G}}}{K_1}$$

Tank A contains 4 m depth of water (39.2 kPa head). Tank B contains 6 m depth of water (58.8 kPa head). There is 90 m of 2.54 cm pipe (205 kPa pressure drop at 75 lpm) and an orifice plate that has a pressure drop of 25 kPa (60% presssure recovery) at a flow rate of 75 lpm. A control valve is required by the process designers to adjust the flow over the range of 10–75 lpm. If the control system designer selects a 1.9 cm globe valve with a fully open C_v of 8.7 and a rangeability of 50:1, then the maximum flow will be approximately 8.7 × [sqr(41/1.0)] / 11.56 = 4.819 m^3/hr = 80.3 lpm. The value of ΔP = 41 kPa = 351 – 310 is found from Figure 5.4 at 81 lpm by subtracting the value on curve *F* from the value on curve *C*. These values were found with a few trials of flow and pressure in the equation and the graph of Figure 5.4. The minimum flow will be approximately (8.7/50) × [sqr(505/1)] / 11.56 = 0.3382 m^3/hr = 5.64 lpm. Notice that the applied flow adjustability is only 81:5.6 or 14.5:1, although the valve C_v ranges from 8.7 to 0.174, or 50:1. The selected valve will satisfy the design requirements of 10–75 lpm adjustability.

Capacity Correction for Piping Geometry

The capacity of a valve can change depending on the abruptness of change in the fluid streamlines entering and leaving the valve. This effect is designated F_p, for Piping Geometry Factor in the ISA *Standard SP75.01: Control Valve Sizing Equations* (Instrument Society of America, 1977).

This reference book gives an approximate table of values for F_p shown in Table 5.1, where *d* is the nominal valve pipe size in inches and *D* is the inside diameter of the pipe in inches before an abrupt contraction to the valve pipe size. This factor can reduce the capacity of the valve significantly.

Control Valve Rangeability

The manufacturer guarantees a range of smooth adjustment of the valve C_v, from a minimum to a maximum value. The maximum value is the rated C_v of the wide open valve. The minimum value is defined as a fraction of the rated C_v. A rangeability ratio of 50:1 means that the minimum value of guaranteed smoothly adjustable C_v is equal to the rated wide open C_v divided by 50. The most commonly available rangeability value is 50:1, but some valves are specified with a rangeability of 35:1. In order to guarantee this value, the manufacturer builds in some extra capability so that the

TABLE 5.1
Approximate table of values for piping geometry factor

	C_v/d^2					
d/D	10	15	20	25	30	80
0.67	0.98	0.95	0.91	0.87	0.83	0.48
0.50	0.96	0.91	0.85	0.79	0.73	0.38
0.25	0.93	0.87	0.79	0.72	0.65	0.31

Source: *Standard SP75.01: Control Valve Sizing Equations.* Reprinted with the permission of the Instrument Society of America (ISA). Copyright © 1977 by the Instrument Society of America.

valve will actually adjust smoothly down to a value less than the rated C_v divided by the rangeability. To find the maximum value of flow, use the rated C_v, and to find the minimum value of flow use the rated value of C_v divided by the rangeability.

Rangeability is not the range of flow that the valve will adjust smoothly. It is the range of valve C_v over which the valve can be adjusted smoothly, from a minimum C_v to the rated C_v. If the pressure drop across the valve remained constant as the flow through it varied, then the rangeability in flow would be the same as the rangeability in C_v. However, for most installations, as the valve closes and the flow decreases, the pressure drop across the valve increases significantly. This means that at low values of flow, there is more pressure forcing fluid through the valve. Therefore the minimum value of flow relative to the maximum value of flow is greater than the minimum value of C_v relative to the maximum value of C_v.

Control Valve Characteristic

The control valve characteristic describes the way the C_v changes relative to the rate of change of the valve control signal. For example, a linear characteristic indicates that the C_v changes linearly with the control signal. Therefore a 1% increase in the control signal for a linear valve with a C_v of 100 changes the valve C_{vp} when at 67.5 to 68.5 or at 2.96 to 3.96. An equal percentage characteristic indicates that the percentage change in C_v is proportional to the change in the control signal. So, a 1% increase in the control signal for an equal percentage valve, with a C_v of 100 and a rangeability of 50, changes the valve C_v when the signal is 90% and the partial C_{vp} is 67.6 to a partial C_v of 70.2 (3.91% increase = 1% of ln 50). When the signal is at 10% and the partial C_{vp} at 2.96, it increases to 3.075 (3.91% increase) for a 1% increase in the signal. The formula for an equal percentage valve that relates the partially open valve C_{vp} to the fully open valve C_{vf} is

$$C_{vp} = \frac{C_{vf}}{R} e^{\frac{y}{y_m}(\ln R)}$$

where C_{vp} is the partially open C_v, C_{vf} the fully open C_v, R the rangeability, e the base of natural logarithms, y the actual signal value, y_m the maximum signal value.

EXAMPLE 5.2 As an example, an equal percentage control valve with a fully open C_v of 106, a rangeability of 50:1, and a signal of 25% has a partially open C_v of

$$5.637 = \frac{106}{50} e^{\frac{25}{100}(\ln 50)}$$

A wider rangeability is achieved by using two valves in parallel.*

Selection of Liquid Flow Adjusting Device

The most common liquid flow adjusting device is the control valve; the variable speed pump and variable volume fixed speed pump are much less common. Appendix B contains photos and diagrams showing the relative sizes of standard control valves, the mounting methods, and the connections to the pipes and vessels.†

The fluid characteristics that may be required to select a control valve include:

- Type of fluid
- Specific gravity of fluid (minimum, maximum, normal values)
- Upstream fluid pressure (minimum, maximum, normal values)
- Downstream fluid pressure (minimum, maximum, normal values)
- Temperature of fluid (minimum, maximum, normal values)
- Flow rate (minimum, maximum, normal values)

The control characteristics that may be required to select a control valve include:

- C_v, flow capacity factor of the valve when wide open.
- Control signal (e.g., 4–20 ma DC or 3–15 psig).
- Direct or reverse actuation requirement.
- Fail open or fail closed requirement.
- Valve stroke or travel, stem displacement in inches or cm from fully closed to fully open (e.g., ¾″). Usually, the valve stroke is linear with the control signal. However, it is possible to obtain a positioner with a characterizing cam that will provide any reasonable nonlinear characteristic relating the valve stem displacement to the control signal.

---

*More information on this is found in a discussion about sequencing-control valves in Anderson (1980).

†Photographs and diagrams are also in Kirk and Rimboi (1975), Anderson (1980), Liptak and Venczel (1985), and Fisher Controls Company (1977).

- Characteristic curve relating the rate of change of valve internal trim opening area to the rate of change of the control signal (e.g., linear or equal percentage characteristic).
- Repeatability, normal variation in stem position when repeating the same control signal (e.g., 0.15% of stroke).
- Rangeability, the ratio (guaranteed by the manufacturer) of maximum smoothly adjustable flow rate to minimum smoothly adjustable flow rate with constant pressure drop across the valve (e.g., 50:1), or the ratio of maximum C_v to minimum C_v.
- Resolution, the minimum possible size step in stem position caused by the corresponding small step in the control signal.
- Frequency response, values of gain (i.e., amplitude of flow rate sine wave divided by amplitude of control signal sine wave), and values of phase shift (i.e., trigonometric degrees that the flow sine wave lags behind the control signal sine wave) over a wide range of sine wave frequencies.
- Closed valve leakage classification.
- Positioner, a servomechanism that forces the valve stem to follow the control signal, may be required.

The mechanical characteristics that may be required to select a control valve include:

- Valve body style (globe, cage, butterfly, and ball)
- Process piping connections (e.g., 2 NPT screwed, flanged, or welded)
- Mounting methods (e.g., pipe mounting)
- Stem operating power (e.g., 20 psig air)
- Acceptable valve noise
- Materials of construction to meet environmental conditions, such as cast iron, stainless steel, bronze, or other materials to meet extremes of temperature, pressure, corrosion, and pressure drop
- Quality-control specifications (e.g., ISO 9000 series)

The valve user specifies the flow requirements, rangeability, control features, and environmental conditions of temperature, pressure, noise, mounting style, and corrosion. The manufacturer of the valve selects the valve body and a corresponding actuator that will meet the user's requirements for control signal (e.g., 4–20 ma DC), operating power (e.g., 115 V 60 Hz), and failsafe conditions (e.g., open on power or control signal failure).

The most common control valve bodies use either in-line motion (globe or cage type) or rotary motion (ball or disc type). The in-line motion valve has a plug that moves in a straight line down into a seat restricting the area of opening and cutting off the flow. For a globe valve the flow rises up through the seat and meets the globular plug and flows around it. For a cage valve the flow also rises up through the seat and meets the plug, which may be partway down inside the cage, cutting off the flow through the slots of the cage cylinder. The cage valve provides superior guidance for the plug compared to the globe valve. The semi-ball or disc (butterfly) of the rotary motion valve allows most of the opening through the pipe to be available to the flow when the valve is open, but the opening is gradually cut off as the disc or semi-ball rotates closed. Usually the rotary motion valve provides greater capacity (for the same size pipe) than the in-line motion valve, and as a result it is cheaper for similar flow rates. It also usually provides satisfactory rangeability and resolution.

The most common control valve actuator is the pneumatic diaphragm with return spring. It is used for in-line and rotary motion. It is fast acting, reliable, safe in explosive environments, and inexpensive. Air pressure forces the diaphragm to move the valve stem in one direction against the spring. With the air pressure reduced or removed, the spring moves the stem against the diaphragm force in the return direction. The force of the diaphragm due to the air is always balanced against the spring force, which depends on the distance that it has been compressed. Usually this actuator is sized to stroke the valve stem from fully closed to fully open for a pneumatic pressure change from 3–15 psig. Although it is sometimes necessary to reverse this action (15–3 psig) so that the valve opens on signal failure.

Other types of actuators include the pneumatic piston and cylinder type, the electric motor type, and the electrohydraulic type. These actuators require positioners to detect the position of the valve stem and to cause the actuator to move the stem in the correct direction. These actuators use a double-acting piston or a reversing electric motor to change the direction of operation. The pistons are fast and very forceful. The electric motor is fairly slow. But it develops much force through a worm gear. The positioner, in its own local control loop, detects the position of the stem, compares this position to the control signal, and when they are equivalent, it holds the piston or motor shaft right at that position.

5.2 GAS FLOW ADJUSTMENT

Adjustment of gas flow closely follows those principles explained in section 5.1, Liquid Flow Adjustment. It is defined as the capability to adjust smoothly the rate of transfer of a gas over a range of transfer rates from a minimum (e.g., 0.1 kg per second) to a maximum (e.g., 1.4 kg per second). The purposes of adjusting the flow of a gas are:

- To set the rate of raw material flowing into a process, such as hydrogen to a margarine making process.
- To set the rate of energy entering a process, such as natural gas to a cement kiln.
- To set the rate of withdrawal of energy from a process, such as cooling air to a cooling tower.
- To set the rate of withdrawal of material from a process, such as compressed air from a compressor.

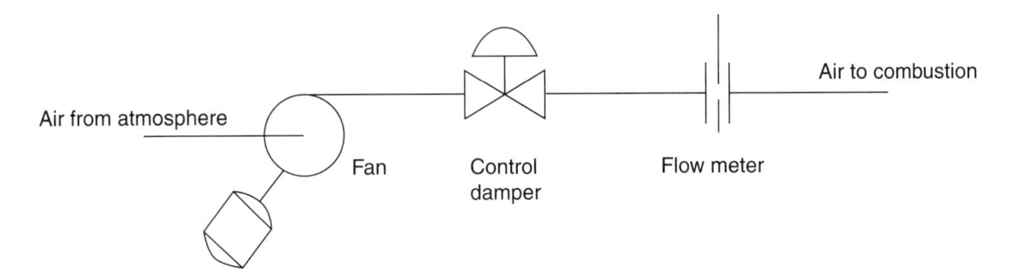

FIGURE 5.5
Typical gas flow process

Gas Flow Process

Figure 5.5 shows a typical gas flow process. Air from the atmosphere is sucked into a fan, then the fan forces the air through a damper, then through a flowmeter, and then into a combustion process such as the burner in a cement kiln. Then most of the air is transformed into other gases such as carbon dioxide and carbon monoxide and they flow along the kiln, up the raw-meal heater, and out the stack. In order to understand this flow process, the fluid pressure drops due to friction, and pressure rise due to the fan, must be related to the flow rate. The pressure drops are like voltage drops in an electrical circuit, and, as in an electrical circuit where the voltage drop increases with increasing current, the pressure drop increases with increasing flow. The pressure at the entrance to the fan is equal to the atmospheric pressure. The pressure rise due to the fan (at full speed it follows the curve *S*) can be found from the fan curve on Figure 5.6, if the flow rate is known or assumed.

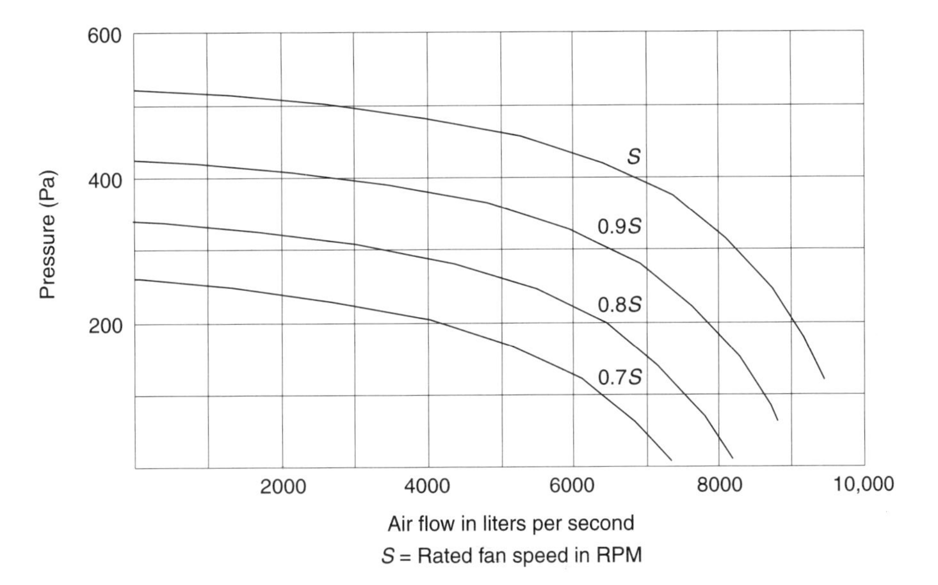

FIGURE 5.6
Fan pressure vs. flow curves

The practical ranges of flow rates in the ducts almost always have a Reynold's number greater than 10,000 and are in the turbulent range, which cause frictional pressure drops that are proportional to the square of the flow rate. (This is the function used for curve *F* in Figure 5.7). Also, the longer the duct the greater the pressure drop; in this case it is directly proportional to the length.* Figure 5.7 shows the pressure drop that the control valve or damper must take up at any particular flow rate, *Z*. The inlet absolute pressure to the fan is P_a due to the atmosphere. The fan pressure rise curve, *S*, from Figure 5.6 does not have P_a added to it to form the supply frictional pressure loss in the piping. The process also does not have P_a added to it to form the demand pressure curve *F*. Again it is only the gauge pressure drop, but it must include the pressure drop in the combustion process and stack, thus returning the gas to the atmosphere in a cycle. To achieve any flow shown on the X axis of Figure 5.7, the control damper must be able to absorb the pressure drop between curve *C* and curve *F* at that flow rate by adjusting its opening to the appropriate value. As shown on Figure 5.7, it can only provide the adjustment between the minimum flow, *N*, and maximum flow, *X*. Since the density of the gas varies with the absolute pressure, it must be taken into account when using the mass flow rate.

*Actual pressure drop in a duct or pipe may be calculated from data provided in Crane Company (1957) and Fisher Controls Company (1977) in the chapter references.

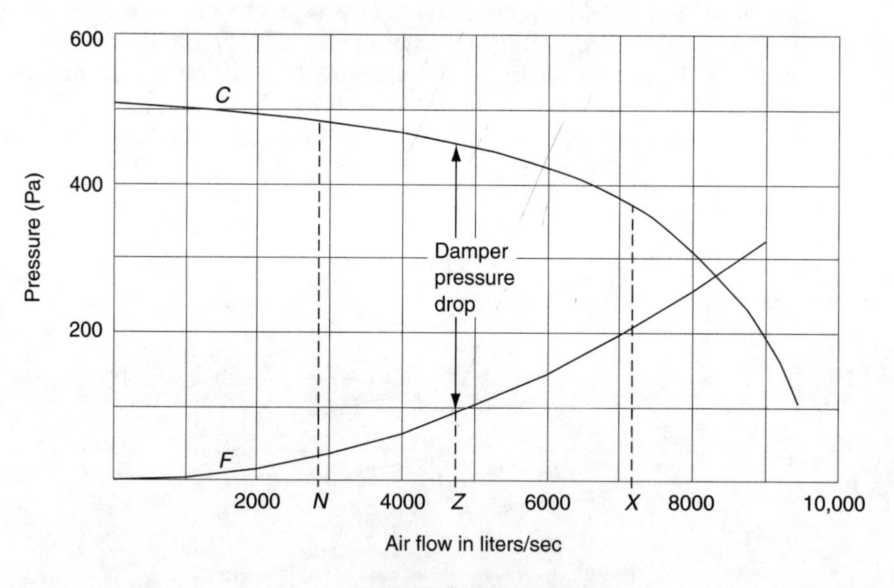

FIGURE 5.7
Pressure vs. flow for typical gas flow process

Capacity of a Gas Control Valve

For gases, the following ISA equation* relates C_v to Q, volumetric flow rate through the valve; M, molecular weight of the gas; T, absolute temperature of the gas at the valve inlet; Z, gas compressibility factor; P_1, absolute pressure at valve inlet; ΔP, valve pressure drop; K_6, constant to reconcile units; and Y, the gas volume expansion factor due to pressure reduction:

$$C_v = Q \times \frac{\sqrt{\dfrac{M \times T \times Z}{P_1 \times \Delta P}}}{K_6 \times Y} \tag{5.3}$$

where:

English Units	*SI Units*
$K_6 = 7320$	$K_6 = 22.4$
Q in standard cubic feet per hour (SCFH) (14.73 psia, 60°F)	Q in m^3/hour (101.3 kPa, 15°C)
T in degrees R	T in degrees K
P_1 in psia	P_1 in kPa absolute
ΔP in psi	ΔP in kPa

M = molecular weight (air = 28.97), may be replaced by $28.97 \times G$, where G is the specific gravity (air has $G = 1.0$).

Y = dimensionless volumetric expansion factor that varies from a value of 1.0 at $\Delta P = 0$, to 0.667 for $\Delta P = 50\%$ of P_1 (critical or choke flow).

Z = dimensionless compressibility factor (use 1.0 to start).

For the mass flow rate, W, the ISA equation (5.4) is:

$$C_v = W \times \frac{\sqrt{\dfrac{T \times Z}{M \times P_1 \times \Delta P}}}{K_5 \times Y} \tag{5.4}$$

where:

English Units	*SI Units*
$K_5 = 19.3$	$K_5 = 0.948$
W in pounds per hour	W in kilograms per hour

For steam and vapors the preferred ISA equation (5.5) is

$$C_v = \frac{W}{K_3 \times Y \times \sqrt{\gamma \times \Delta P}} \tag{5.5}$$

*This equation is a simplification of the one that appears in *Standard SP75.01: Control Valve Sizing Equations* (Instrument Society of America, 1977).

where:

English Units	*SI Units*
$K_3 = 63.3$	$K_3 = 2.73$
W in pounds per hour	W in kilograms per hour
γ = specific weight of inlet vapor (lb/ft^3)	γ = inlet specific weight (kg/m^3)

The numerical value of C_v for gas should equal the C_v for liquid for the same valve. As for liquid, the C_v is usually considered to be the capacity of the valve when it is fully open, which is expressed as C_{vf}. However, when the valve is partially open, a smaller value, C_{vp}, can be used in the same equation for the reduced capacity.

Choke Flow in a Gas Control Valve If the absolute value of the downstream pressure on a valve is less than one half the absolute value of the upstream pressure, then the velocity of gas or steam flowing through the valve approaches sonic velocity, also known as critical velocity and choke flow, which it cannot exceed, even if the downstream pressure should decrease further. For gas or steam or vapor flow calculations or valve sizing, use a maximium value for ΔP of $P_1 - (P_1)/2 = (P_1)/2$.

Y, the gas expansion factor, varies linearly from a value of 1.0 at zero pressure drop in the valve to a value of 0.667 at the pressure drop (corrresponding to critical flow) that is 50% of the inlet absolute pressure. Start the calculation with $Y = 1.0$.

Z, the compressibility factor, is used to compensate for gases that deviate significantly from the law that relates pressure, volume, temperature, and specific weight for ideal gases. Values of this factor may be found in Hutchison (1976).

EXAMPLE 5.3 **Control Valve for Natural Gas in 4″ Pipe**

Assume a control valve is required to adjust natural gas flowing in a 4″ pipe where the normal upstream pressure is 350 psia, the normal flow rate is 600,000 SCFH (standard cubic feet per hour), the acceptable pressure drop at normal flow is 50 psi, and the gas temperature ranges from 40–100°F. The maximum flow rate anticipated is 800,000 SCFH; and the valve must be able to adjust the flow down to 100,000 SCFH under the worst conditions. What is the smallest size valve that will perform satisfactorily, and what are its minimum value of flow when wide open and maximum value of flow when it is closed to its minimum rangeability (50:1)?

Using ISA equation 5.3, set $Z = 1$, $Y = 1$, $M = 17.5$ for natural gas:

$$\text{Maximum } C_v \text{ required} = 800{,}000 \times \frac{\sqrt{\dfrac{17.5 \times (100 + 460) \times 1}{350 \times 50}}}{7320 \times 1}$$

$$= 81.78$$

If a 2½″ globe valve is selected with a $C_v = 86.5$ and a rangeability of 50:1, then the least flow when the valve is wide open is

$$Q = 800,000 \times \frac{86.5}{81.78}$$
$$= 846,172 \text{ SCFH}$$

The maximum flow (critical or choke flow) when the valve is at minimum adjustable opening is

$$Q = \frac{86.5}{50} \times 7320 \times \sqrt{\frac{\frac{350 \times 350}{2}}{17.5 \times (40 + 460)}}$$
$$= 33,505 \text{ SCFH}$$

This valve is expected to provide satisfactory adjustment of this process over the range of 100,000 to 800,000 SCFH.

EXAMPLE 5.4 **Control Damper for Combustion Air**

As an example using Figure 5.7 and control valve equation 5.3, combustion air is specified at 20°C with a maximum flow requirement of 7000 liters per second (lps), using an orifice plate that has a pressure drop of 25 Pa (60% pressure recovery) at a flow rate of 8000 lps. If we select a 76 cm butterfly damper with a C_v of 32,000 and a rangeability of 30:1, then the maximum flow will be approximately 32,000 × 22.4 × 1 × sqr [101 × 0.120 / (28.97 × 293 × 1)] = 27,085 m^3/hr = 7523 lps, where Y is 1.0, P_1 is 101 kPa, ΔP is 0.120 kPa, M is 28.97, T is 20 + 273°C, Z is 1.0. These values were found with a few trials using the graph of Figure 5.7. The minimum adjustable flow will be approximately (32,000 / 30) × 22.4 × sqr [101 × 0.505 / (28.97 × 293 × 1)] = 1852 m^3/hr = 514 lps. Notice the applied flow adjustability is only 7523:514 or 14.6:1, although the valve C_v ranges from 1067 to 32,000, or 30:1.

Control Valves, Dampers, and Louvres

The most common gas flow adjusting device is the control valve, damper, or louvre. The variable speed fan is much less common. The photos in Appendix B show the relative sizes of these standard control valves, the mounting methods, and the connections to the pipes and vessels.*

The capacity correction factor for piping geometry applies to gas flow control valves in the same way as for liquid flow control valves as described on pp. 112–113. The rangeability and the characteristic of gas control valves are specified in exactly the same way as for liquid control valves. Rangeability is described on p. 113. The characteristic is described on pp. 113–114.

The features used to select gas control valves are similar to those used to select liquid control valves and are listed on pp. 114–116. The damper is very much like a

---

*For additional illustrations, see Kirk and Rimboi (1975), Anderson (1980), Liptak and Venczel (1985), and Fisher Controls Company (1977).

butterfly valve, and the louvre is like a series of rectangular butterflies mounted above each other in a frame so that they overlap one another a small amount when they are closed. Generally the damper and louvre operate in a duct at low pressures less than 100 kPa gauge and usually less than 10 kPa gauge.

Variable Speed Fans

Upon examination of Figure 5.6 we see that the fan pressure vs. flow curve produces less pressure as the speed of the fan is reduced at any given flow rate. Generally, the fan pressure is proportional to the square of the speed. With a variable speed fan, it is possible to control gas flow so that no control valve or damper is required, and for large fans this is becoming very desirable as compared to a fixed-speed fan running at full speed with a damper throttling the flow and throwing away power.

In Figure 5.7 the point where curve *C* meets curve *F* specifies the flow that will result if no control valve exists. If we can slide curve *C* up or down by adjusting the speed of the fan, then this point changes over a wide range and gives us flow adjustment equivalent to control valve adjustment, but without a control valve.

PROBLEMS AND LAB ASSIGNMENTS

5.1 A variable speed pump is selected to transfer liquid from Tank A to process B. A control valve is not included in this line. Instead, a vertical section of pipe is placed downstream of the pump. This piece of pipe rises vertically to just over the height of Tank A, and then back down to the pump elevation. When the pump motor is stopped, the liquid in Tank A will not be able to windmill the pump and drive liquid (water) up the pipe and into process B. The height of the vertical pipe is 7 m, the normal level of liquid in Tank A is 4 m. The demand pressure (kPa) curve of the process is 50, 60, 70, 85, 115, 155, 190, 250, 310, and 375 at 0, 100, 200, 300, 400, 500, 600, 700, 800, and 900 lpm. The supply pressure (kPa) curve, not including the effect of the pipe, at full speed, *S,* is 560, 550, 540, 530, 520, 500, 480, 425, 360, and 240 at the corresponding flow rates of the demand pressure curve. The supply pressure at any reduced speed will be a ratio of the square of the reduced speed to the square of the full speed times the pressure at full speed, and it will occur at the flow rate of the full speed. Establish the flow that will result at speeds 0.4, 0.6, 0.8, and 1.0 of full speed.

5.2 What valve C_v is required for a maximum flow of 160 GPM of water between 40°F and 140°F with an available pressure drop of 14 psid for the valve? As the water temperature changes, how much does the required C_v change? What is the approximate valve size required in inches: ½″ has $C_v = 2.5$, ¾″ has 8.6, 1″ has 20, 1½″ has 34, 2″ has 60, 2½″ has 87, 3″ has 130?*

*Sizes are for a single-ported globe valve described in Fisher Controls Company (1977), pp. 74–75.

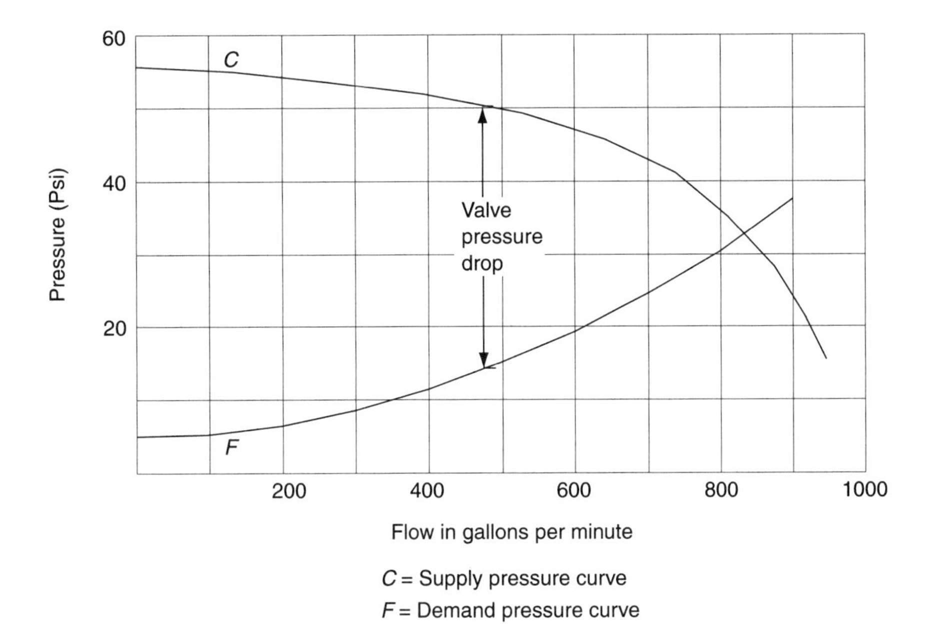

FIGURE 5.8
Graph for problem 5.3

5.3 A Fisher Controls 3″ Rotary V-notch ball valve ($C_v = 372$) is available (rangeability of 50 to 1) to control gasoline flow in a pipe. The specific gravity of the gasoline is 0.83. The supply pressure curve and the demand pressure curve are shown in Figure 5.8. What is the maximum flow and the minimum flow that can be adjusted smoothly with this control system?

5.4 Some standard available control valves are as follows:

Pipe Size	Fully Open C_v
½″	5
1″	17
1½″	34
2″	60

Calculate the fully open C_v required and select an appropriate control valve from these standard valves to maintain water in a tank at a set level. The control valve will be placed in the inlet pipe to the tank. The maximum rate that the water will be drawn out of the tank will be 140 GPM. At 140 GPM flowing in the inlet pipe, the pressure just upstream of the control valve will be 25 psig. The pressure just downstream of the valve will be atmospheric.

5.5 A 1″ control valve from the standard valves listed in Problem 5.4 empties water into a tank. The valve can smoothly adjust its C_v from its fully open value down to 1/50th of its fully open value. If the pressure available at the minimum flow rate is 64 psig, what is the value of the minimum flow in GPM that can be smoothly adjusted?

5.6 A 1″ equal-percentage, cage-guided control valve with a fully open C_v of 20 and a rangeability of 35:1 is specified for control of gasoline flowing in a pipe. The specific gravity of the gasoline is 0.81. The supply pressure curve, *C*, and the demand pressure curve, *F*, are shown in Figure 5.9. What are the maximum flow and the minimum flow that we are sure can be adjusted smoothly with this control valve?

5.7 For the valve of problem 5.6 what is the partially open C_v when the signal to the valve is 10.8 ma if the signal range is 4–20 ma corresponding to $C_{vp} = 0.571$ to 20?

5.8 A control valve adjusts water flowing from Tank A to Tank B as shown in Figure 5.10. This control valve has a characteristic curve relating C_v to the percent of valve opening as shown in Figure 5.11. Use Figure 5.12 to find the flow at valve openings of 5, 10, 20, 40, 60, and 100%. Then plot your values of flow versus the percent of valve opening on Figure 5.11 using your own vertical scale for flow. Mark your flow scale on the right hand side of Figure 5.11.

5.9 What is the adjusted pressure rise for the fan of Figure 5.6 on p. 117 at a flow rate of 5000 lps, if the fan rated speed is 1750 RPM and the adjusted speed is 1365 RPM?

5.10 For the process control system of Example 5.4, what will the maximum and minimum flows be if the fan runs at 70% of rated speed?

5.11 What size control valve is required for a steam jacket using 40 psig saturated steam at 1000 lb/hour if the jacket operates at atmospheric pressure? Choose valve size from sizes given for problem 5.2.

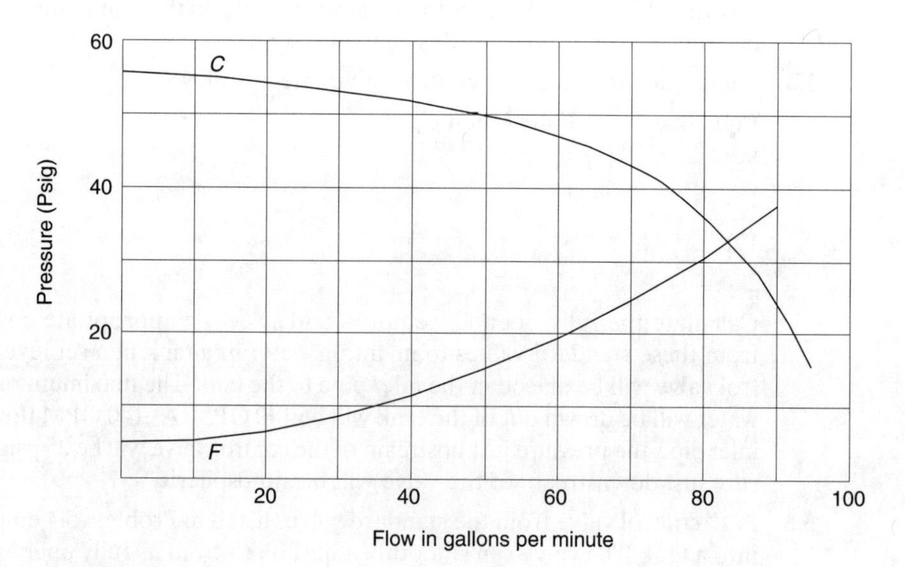

FIGURE 5.9
Graph for problem 5.6

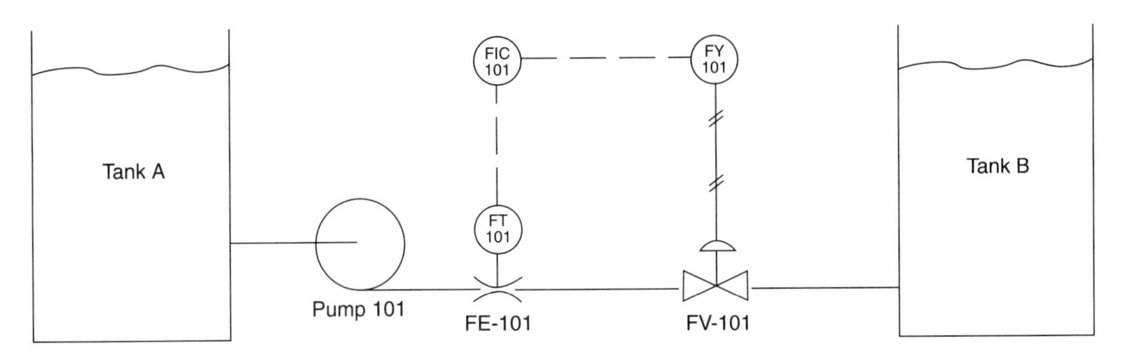

FIGURE 5.10
Diagram for problem 5.8

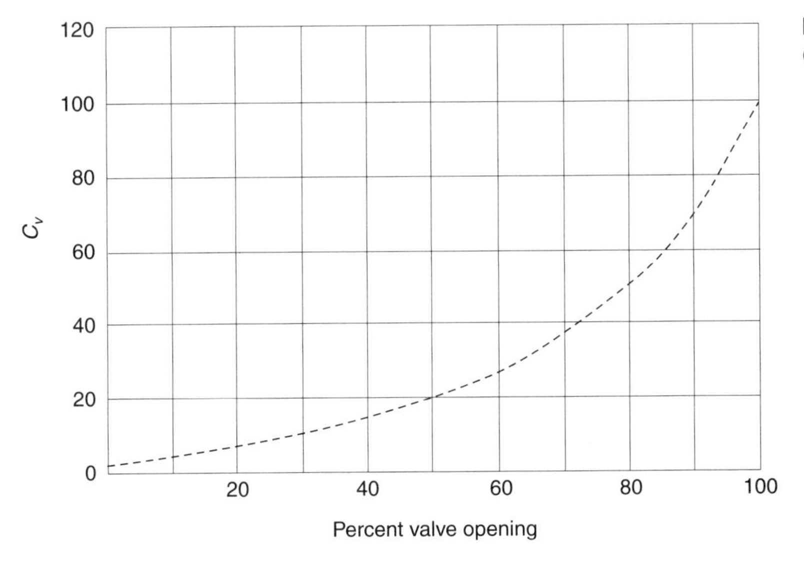

FIGURE 5.11
Graph for problem 5.8

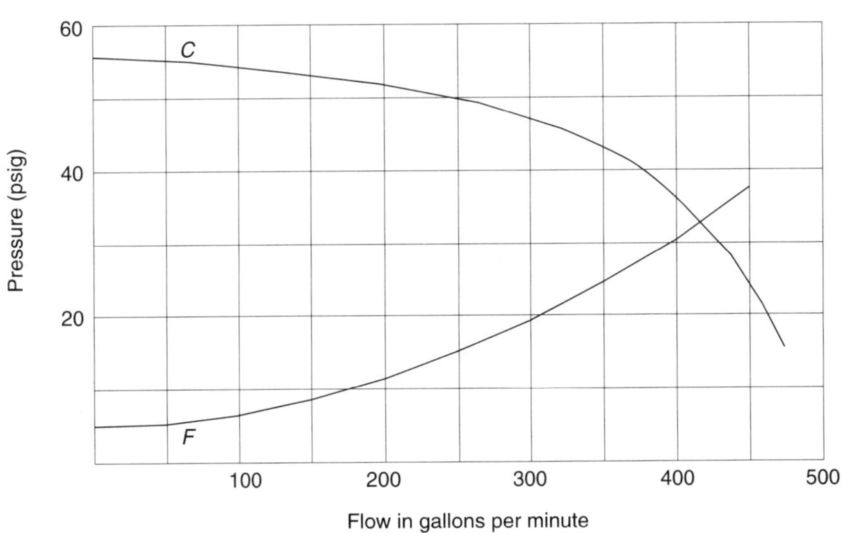

FIGURE 5.12
Graph for problem 5.8

5.12 A butterfly valve is required to adjust the flow of air to a furnace. In order to burn all the fuel at the maximum firing rate, a maximum flow of 20,000 SCFH is required. At the minimum firing rate, an air flow of 2000 SCFH is required. At the maximum firing rate, the available pressure drop across the valve is 0.04 psi. At the minimum firing rate the valve must take up a pressure drop of 0.2 psi. The air is at a temperature of 60°F and at atmospheric pressure. Choose the smallest-sized valve that will provide the required adjustment in air flow from the following valves, each of which has a rangeability of 35 to 1:

Valve Pipe Size	Valve C_v
2″	100
2½″	150
3″	250
4″	500
6″	1500

What will the maximum flow rate be with the valve that you have chosen? What will the minimum flow rate be with the valve that you have chosen?

5.13 The fan of Figure 5.6 provides the supply pressure curves for the process of Figure 5.7. There is no control damper in the duct. Estimate the minimum and maximum flow rates that will be provided if the fan can be adjusted from $0.5 \times S$ to $1.05 \times S$.

5.14 **C_v of Valve for Liquids**

Objective: To measure the C_v of a control valve for liquids at approximately 10, 30, 50, 70, 90, and 100% of valve stem position.

Equipment:
- Small (⅛″ to ½″) pneumatic (3–15 psig) control valve [Use the same valve for liquid and gas (project 5.15).]
- Adjustable (0–20 psig) pnematic supply for valve signal
- Small (approximately 8″ to 15″ diameter, 36″ high) water tower with level gauge
- Calibrated container, 0 to 5 gallons
- Stopwatch

Use Procedure 1 or Procedure 2.

Procedure 1—Valve on tower exit: Connect the valve inlet to the bottom of the tower so that water from the tower will flow through the valve into a hose that can be flipped quickly from the drain into the calibrated container and back again. Maintain the water level in the tower at 27.7 in. (1 psi) while it flows through the valve.

Make a table with columns of Percent Open, Signal Pressure (psi), Level (inches), Pressure Drop (psi), Quantity (gallons), Elapsed Time (seconds), Elapsed Time (minutes), Flow (GPM), and C_v. Make separate runs at 10, 30, 50, 70, 90, and 100% open and collect data for level in the tower, quantity in the container, time (seconds) elapsed to collect the quantity (take about 1 to 3 minutes for each run). Calculate all the other entries in the table. Plot a graph of C_v (vertical axis) against Percent Open (horizontal axis). Draw the flow diagram showing your process.

Procedure 2—Valve on tower inlet: Connect the valve inlet to a water supply and its outlet to the tower inlet. Provide a pressure gauge on the inlet side of the valve. Measure the tower diameter and establish the number of gallons per inch of height.

Make a table with columns of: Percent Open, Signal Pressure (psi), Tower Level Change (inches), Valve Pressure Drop (psi), Quantity (gallons), Elapsed Time (seconds), Elapsed Time (minutes), Flow (GPM), and C_v. Make separate runs at approximately 10, 30, 50, 70, 90, and 100% valve opening and collect data for valve signal pressure (psig), level change in the tower, quantity (gallons), time (seconds) elapsed to collect the quantity (take about 1 to 3 minutes for each run). Calculate all the other entries in the table. Plot a graph of C_v (vertical axis) against Percent Open (horizontal axis). Draw the flow diagram showing your process.

Conclusions: What is the 100% open C_v of the valve? How does this compare with the nameplate value? What type of characteristic does the valve have? What equation did you use to calculate C_v?

5.15 C_v of Valve for Gases

Objective: To measure the *Cv* of a control valve for gases at 10, 30, 50, 70, 90, and 100% of stroke.

Equipment:
- Small (⅛″ to ½″) pneumatic (3–15 psig) control valve [Use the same valve for liquid (project 5.14) and gas.]
- Adjustable (0-20 psig) air supply for control signal
- Adjustable (0–20 psig) air supply for test fluid
- Air rotameter 0–100 SCFH
- Differential pressure indicator 0–2 psi
- Air pressure gauge 0–15 psig

Procedure: Connect the valve inlet to the adjustable air test fluid so that air will flow through the valve and then through the air rotameter to the atmosphere. Connect the differential pressure indicator so that it measures the pressure drop across the valve. Connect the pressure gauge so that it measures the pressure at the valve outlet. Make a table with columns of: Percent Open, Valve Signal Pressure (psig), Valve Outlet Pressure (psia), Valve Pressure Drop (psi), Flow (SCFH), and C_v.

Make separate runs at 10, 30, 50, 70, 90, and 100% open and collect data for valve outlet pressure, pressure drop, and flow. Calculate the C_v. Plot a graph of C_v (vertical axis) against Percent Open (horizontal axis). Draw the flow diagram showing your process.

Conclusions: What is the 100% open C_v of the valve? What type of characteristic does the valve have? What equation did you use to calculate C_v? How does the graph for air compare with the graph for water?

5.16 Measurement of a Solenoid Valve C_v

Objective: To use air to measure the C_v of an electric solenoid valve.

Procedure: Use an electric solenoid valve and measure its C_v as you did in project 5.15 while it is held open electrically. Connect the solenoid valve inlet to a regulated air supply set at about 10 psig. Connect the outlet of the valve to an air flow rotameter that exhausts to the atmosphere. Connect a differential pressure gauge across the inlet and outlet of the solenoid valve. Read the pressure drop and air flow at several values of air flow. Calculate the C_v at each value of air flow.

Conclusions: Compare the measured C_v to the nameplate value.

REFERENCES

Anderson, N. A. 1980. *Instrumentation for Process Measurement and Control.* 3rd ed. Radnor, PA: Chilton Company.

Bateson, R. N. 1999. *Introduction to Control System Technology.* Upper Saddle River, NJ: Prentice Hall.

Crane Company. 1957. *Flow of Fluids Through Valves, Fittings, and Pipe: Technical Paper 410.* Chicago: Crane Company.

Fisher Controls Company. 1977. *Control Valve Handbook.* Marshalltown, IA: Fisher Controls Company.

Hutchison, J. W., ed. 1976. *ISA Handbook of Control Valves.* 2nd ed. Triangle Park, NC: Instrument Society of America.

Instrument Society of America. 1977. *Standard SP75.01: Control Valve Sizing Equations.* Triangle Park, NC: Instrument Society of America.

Kirk, F. W. and Rimboi, N. R. 1975. *Instrumentation.* 3rd ed. Chicago: American Technical Publishers, Inc.

Liptak, B. G. and Venczel, K. 1985. *Instrument Engineers' Handbook—Process Control.* Rev. ed. Radnor, PA: Chilton Company.

6

AUTOMATIC CONTROL

OBJECTIVES

When you have completed this chapter you will be able to:

- Describe the effect of set-point disturbances to a control loop
- Describe the effect of load disturbances to a control loop
- Describe the desired performance of a control loop
- Describe the effects of loop gain on loop stability
- Describe a control loop with a proportional-only controller
- Describe a control loop with a proportional plus integral controller
- Describe a control loop with a proportional plus derivative controller
- Describe a control loop with a proportional plus integral plus derivative controller
- Perform the ultimate cycle procedure for tuning a control loop
- Calculate the proportional gain to achieve quarter amplitude response
- Calculate the proportional plus integral gains to achieve quarter amplitude response
- Calculate the proportional plus derivative gains to achieve quarter amplitude response
- Calculate the proportional plus integral plus derivative gains to achieve quarter amplitude response
- Describe ratio control
- Describe cascade control

6.1 INTRODUCTION TO THE CONTROLLER OF THE CONTROL LOOP

A typical continuous as opposed to discrete closed control loop is shown in Figure 1.1 on p. 4 and described in sections 1.3 to 1.7 on pp. 3–7. The objectives of a closed control loop are, first, to provide the process operator with a simple method for changing the measured process variable to a desired value (set point) and, second, to maintain the measured process variable at the desired value, even though the process is disturbed when other process variables are changing.

The most important feature of the control loop controller block is the comparison of the measured process variable to the set point. This is done by subtracting the feedback signal from the set-point signal. Generally, 4–20 ma DC is converted to 1 to 5 V by means of a precise 250 Ω resistor, and the two voltages are placed in series so that their difference is the error signal for the controller. The error signal is used to produce the controller output signal that controls the process adjusting device and, as a consequence, the adjusted process variable. The object is to change the measured process variable so that it becomes closer to the set point value. There are several functions of the error signal that may be used in various combinations to generate the controller output signal. The most common function of the error signal is the proportional controller output signal. In this case the output of the controller is proportional to a constant number, called the *gain,* multiplied by the error signal. All of the standard functions of the error signal include the proportional function, either by itself or with the addition of integral (of error) action or derivative (of error) action or both integral and derivative action. The proportional action alone changes the controller output instantaneously to a new value whenever the error signal changes. The integral and derivative actions are functions of the way the error signal changes with time and they produce added changes to the controller output signal depending on the effects of time on the error.

Each of the controller functions requires tuning in order to achieve the most desirable response of the loop to a disturbance, such as a change in the set point. Tuning requires the setting of a numerical value for each controller function, rather like setting a dial knob for each function. A variety of undesirable responses occur if any of the functions are poorly tuned. These responses are described in the next section.

6.2 LOOP STABILITY

Disturbances

There are two major types of disturbances that affect a closed control loop. The first is due to a set-point change, the second is due to a load change. It is easy to understand how an operator can change a set point, and for the example of level loop 132 in Figure 6.1, when the set point is increased it becomes necessary for the loop to increase the flow into the tank so that it is temporarily more than the flow out until the level reaches the new value of the set point. Then the loop will cut the flow back so that it is again equal to the flow out of the tank. For a short time the level is not equal to the set point. It may even overshoot the set point before settling down at or near the set point. When the level settles down again to a steady value, the disturbance is considered over.

A load change disturbance also may occur to level loop 132. The flow out of the tank is considered the load for this loop. If the loop is running smoothly with the flow in equal to the flow out, and the level is steady at or near to the set point, then the loop is in equilibrium. If the flow out is changed due to an operator changing the outflow

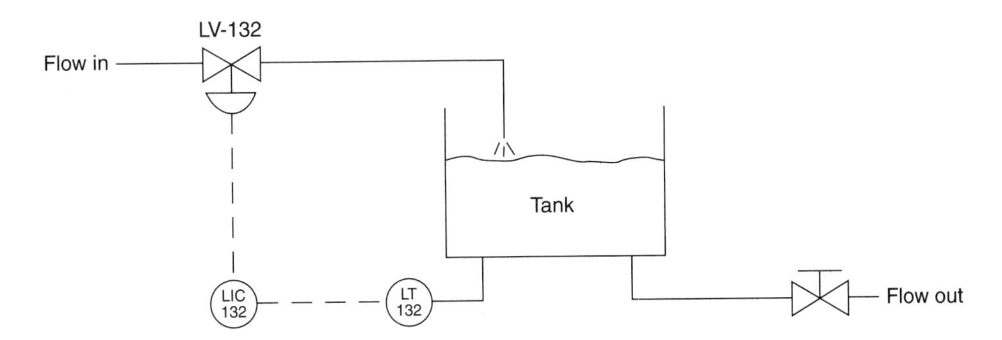

FIGURE 6.1
Level control loop 132

valve or due to the process downstream automatically opening or closing another valve in order to obtain more or less flow, then the loop will be disturbed. For example, assume the flow downstream is reduced by 25%, the level in the tank will start to rise, and after a short time the control loop 132 will start to close the inflow valve LV-132. The level will first rise above the set point and then fall back to the set point, perhaps dropping down below the set point, before settling down to a steady value at or near the set point. If the level oscillates about the set point for several oscillations, then the frequency and the dampening of the oscillations will be the same for either a set point or a load disturbance. The frequency and damping characterize the response of the measured variable to any type of disturbance that affects the loop.

The tuning of the controller gains provides a variety of responses to the disturbances, some of which are in the desirable category, but most are undesirable. As the gain is increased from a very small value, to an intermediate value, and then to a very large value, the response of the measured process variable changes from being sluggish and nonoscillatory, then to more rapid and smooth, and finally to very rapid and oscillatory. This is shown in the diagrams of Figure 6.2, with set-point disturbances shown in the left column and load disturbances in the right column. The dotted line is the set-point value. The full line is the value of the measured process variable.

The responses for the very small, small, and intermediate gains are stable, whereas the responses for large and very large gains are unstable. At time t_1, the disturbance begins. For the set-point disturbance cases, the set point is stepped up to a higher value to cause the disturbance. For the load disturbance cases, the load variable is stepped. Assume that the responses are all stable before t_1. Also assume that the gains are set an instant before t_1.

Desired Performance of a Control Loop

The variety of responses shown in Figure 6.2 indicate that the most desirable responses are in the range from critically damped response to underdamped response. Thus the tuning of the gains must be done to achieve the performance that exists in this range.

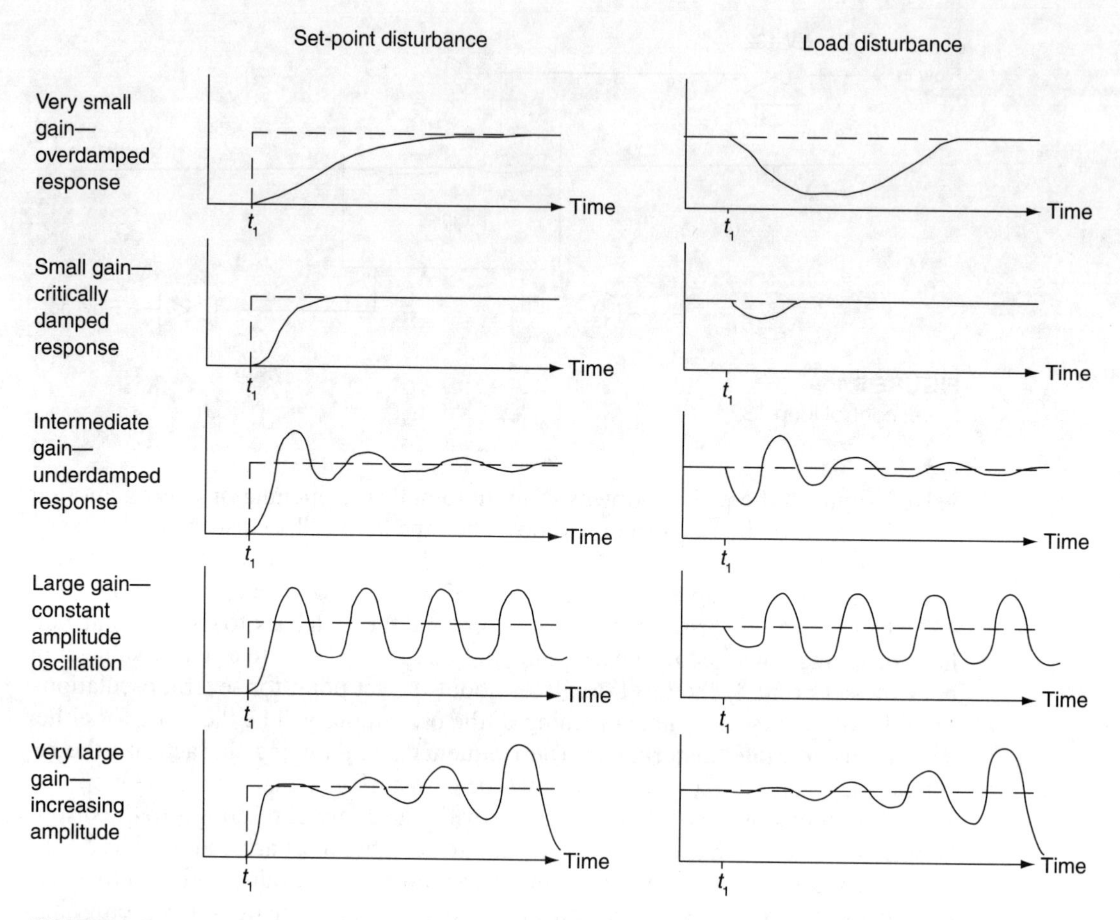

FIGURE 6.2
Types of responses to disturbances as gain is increased (*Source:* R. Bateson. 1995. *Introduction to Control System Technology.* 5th ed. Upper Saddle River, NJ: Prentice Hall. Reprinted by permission of Prentice-Hall, Inc., Upper Saddle River, NJ.)

If we examine the underdamped response to a load disturbance, we will notice three important features of the error. The most desirable performance would have no error. But disturbances always result in some error. First, there is a maximum value of error. This is the maximum distance that the measured process variable is away from the set point during the disturbance. Second, there is the time that the measured process variable takes to settle back to a steady value after the disturbance begins. Finally, there is the steady error that may exist after the process variable settles down. For good performance: the maximum value of the error needs to be as small as possible, the settling time needs to be as short as possible, and the steady error needs to be as small as possible.

Effects of Loop Gain on Loop Stability

The effects of loop gain on stability is summarized in this section. A simple loop with proportional controller, process adjusting device, process, and sensor lumped together in the one block is shown in Figure 6.3. At a certain frequency, for example 1.3 Hz, the overall open loop gain, *K*, of this controlled system equals 1.0 with a phase shift of –180°. Figure 6.4 shows the way the set point, *SP*, and the error, *ERR*, and the feedback,

FIGURE 6.3
Closed loop with gain = *K*, phase shift = –180°, at 1.3 Hz

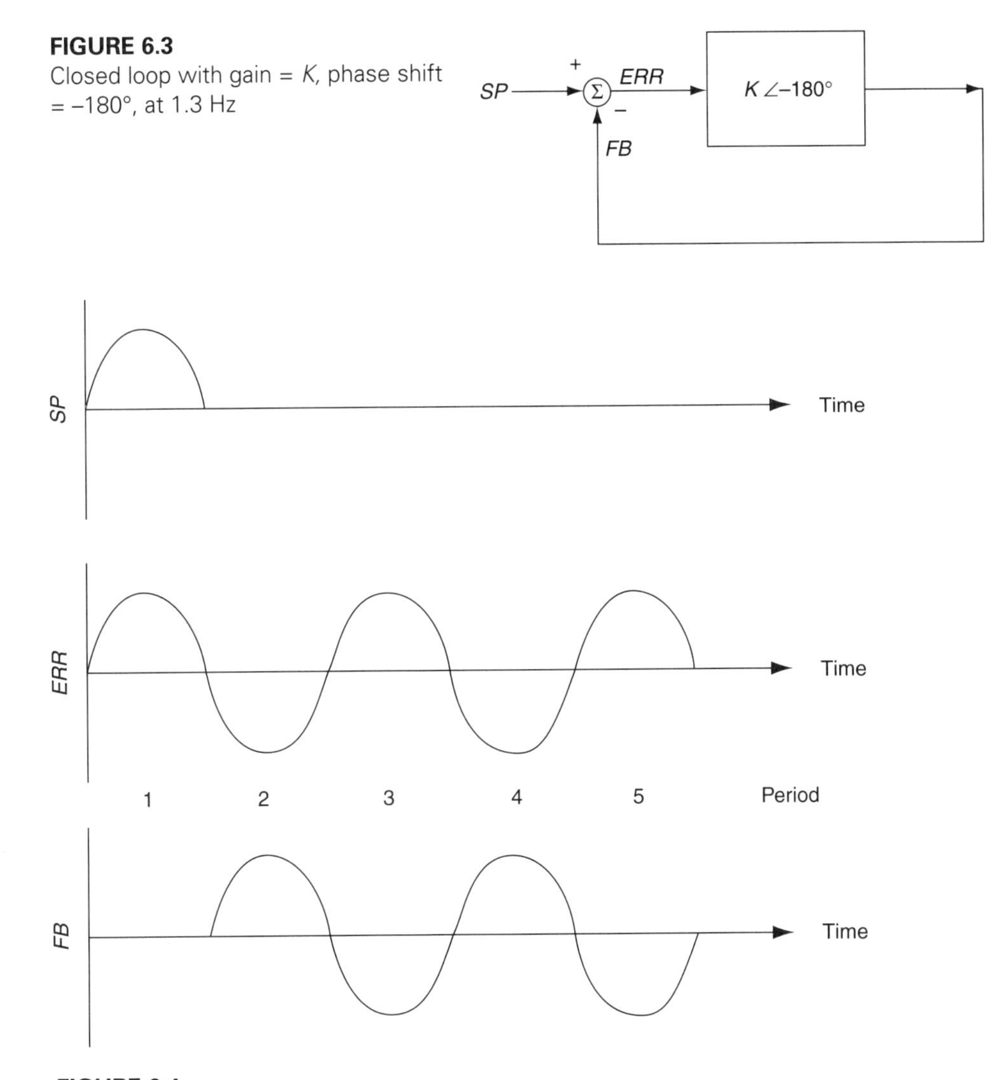

FIGURE 6.4
Constant amplitude oscillations, gain $K = 1$ (*Source:* R. Bateson. 1995. *Introduction to Control System Technology.* 5th ed. Upper Saddle River, NJ: Prentice Hall. Reprinted by permission of Prentice-Hall, Inc., Upper Saddle River, NJ.)

FB, vary as time passes after a half-wave pulse of 1.3 Hz is applied to the set point during period 1. In that period the feedback value is zero, so the error is equal to the set point. The controller output, or *FB,* is phase shifted by –180° from the error into period 2. Since the set point is zero in period 2, then the error is the negative of *FB* in period 2. This value of the error is phase shifted by –180° into period 3 to produce the feedback *FB*. As is shown, a constant amplitude oscillation occurs, and this is considered an unstable response.

If the gain, *K,* in Figure 6.3 is changed to 0.5, the responses of *ERR* and *FB* to the half-wave pulse at the set point are shown in Figure 6.5. This is considered a stable underdamped response with oscillations that are dampened as time passes.

If *K* in Figure 6.3 is changed to 2.0, the responses of *ERR* and *FB* to the half-wave pulse at the set point are shown in Figure 6.6. This is obviously an unstable, increasing amplitude oscillation. We must find the value of gain, *K,* that will make the loop perform in the most desirable way.

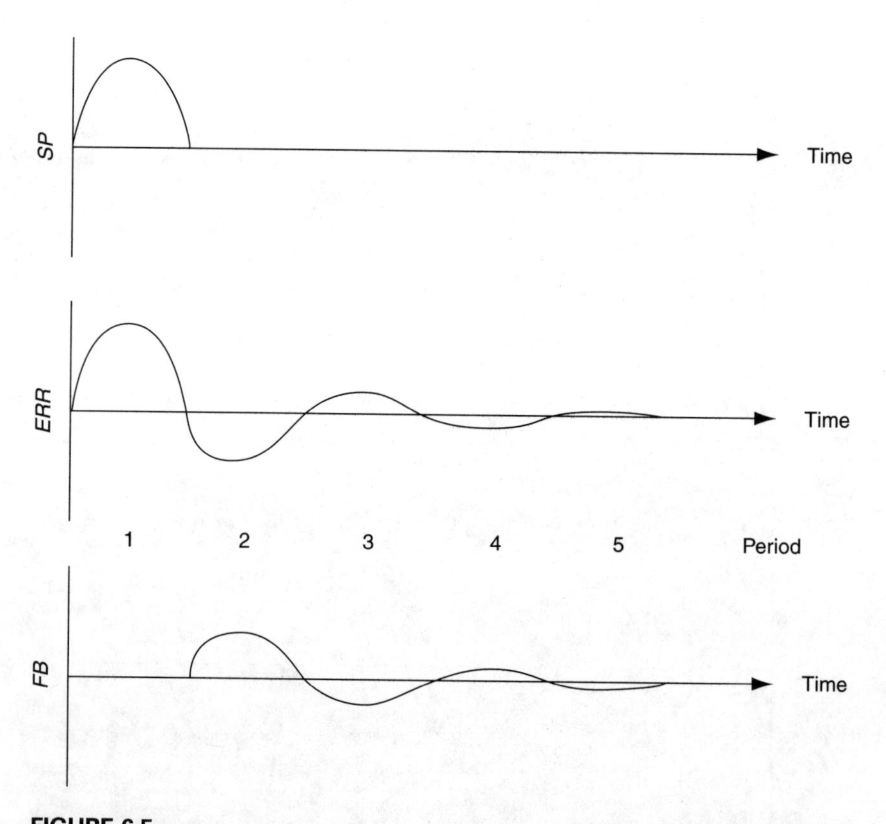

FIGURE 6.5
Underdamped oscillations, gain $K = 0.5$ (*Source:* R. Bateson. 1995. *Introduction to Control System Technology.* 5th ed. Upper Saddle River, NJ: Prentice Hall. Reprinted by permission of Prentice-Hall, Inc., Upper Saddle River, NJ.)

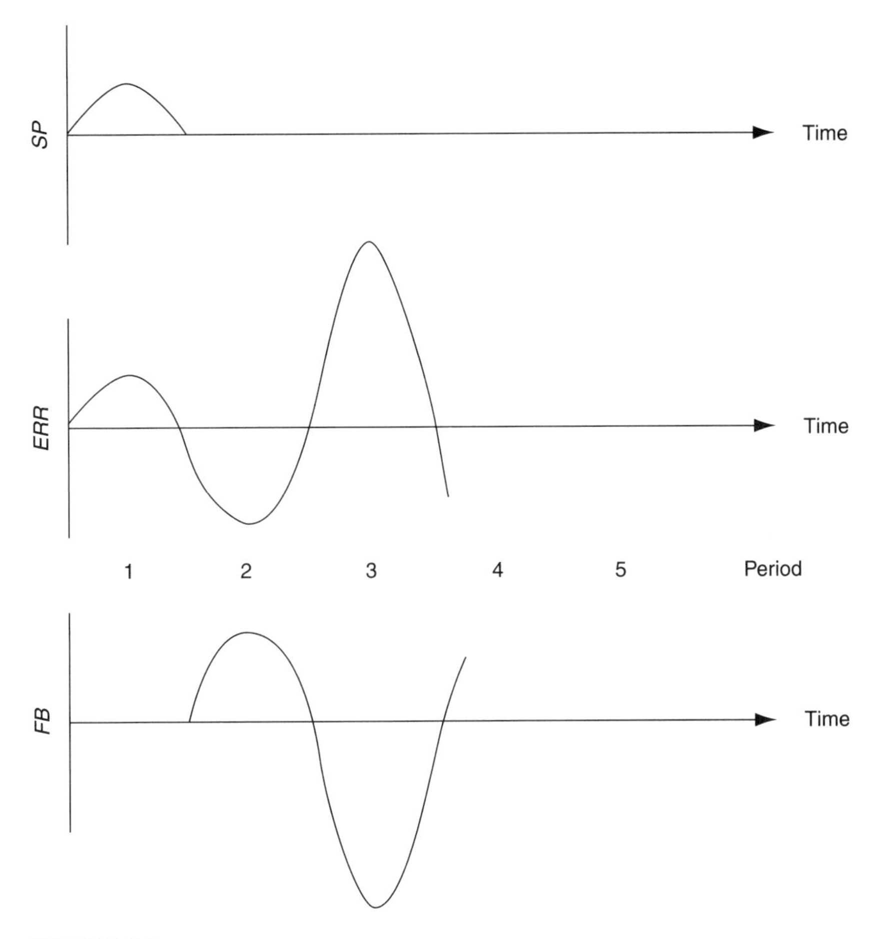

FIGURE 6.6
Increasing amplitude oscillations, gain $K = 2$ (*Source:* R. Bateson. 1995. *Introduction to Control System Technology.* 5th ed. Upper Saddle River, NJ: Prentice Hall. Reprinted by permission of Prentice-Hall, Inc., Upper Saddle River, NJ.)

6.3 PROPORTIONAL CONTROL

Proportional control can be most simply described with an example of level control. In Figure 6.1 the level in the tank is sensed by level transmitter, LT-132, adjusted by level valve, LV-132, and controlled by level controller, LIC-132. Therefore, level loop 132 is a closed loop providing automatic level control of the tank to the set point that has been set into LIC-132 by a human operator. For this loop the measured process variable is the tank liquid level, the feedback signal is the 4–20 ma DC signal from level transmitter LT-132, the process adjusting device is the level control valve, LV-132, and the adjusted process variable is the flow of liquid into the tank. The level controller, LIC-132, is the center of attention in this section of the chapter.

Figure 6.7 shows the faceplate of level controller LIC-132. This shows a typical controller made of electronic hardware that one would see on a control panel. It is often shown like this on a computer screen so that an operator can readily relate to it. The leftmost bar indicates the value of the set point in level units, 0–100% full or 0–2 m. The middle bar indicates the value of the level. (It shows the value of the feedback signal from LT-132.) The rightmost bar shows the controller output signal in percent (0–100). If the manual-auto switch is in manual, then the controller output signal depends only on the setting of the manual adjustment knob. If the switch is in auto, then the controller output signal is set automatically by the controller, depending on the error signal value *(err)*, the gain, K_p, and the bias value, which is internally set inside the controller, usually to 50%, and unavailable to the operator. This may be seen in the block diagram shown in Figure 6.8. The value of the gain, K_p, shown in Figure 6.8, may be adjusted by pulling out the faceplate from the control panel and then adjusting the K_p knob inside the controller.

When the liquid level reaches the value of the set point under proportional control, then the error is zero and the controller output signal is $50\% = 0 \times K_p + 50$, as can be seen from Figure 6.8. But assume the gain, K_p, has been tuned to a value of 2.4 in order to achieve acceptable, stable operation of the loop. Also assume the set point is at 43% and the flow out of the tank is at 59.6%. Now calculate the actual controller error and actual feedback signal required to maintain the loop and the tank in equi-

FIGURE 6.7
Faceplate of level controller LIC-132

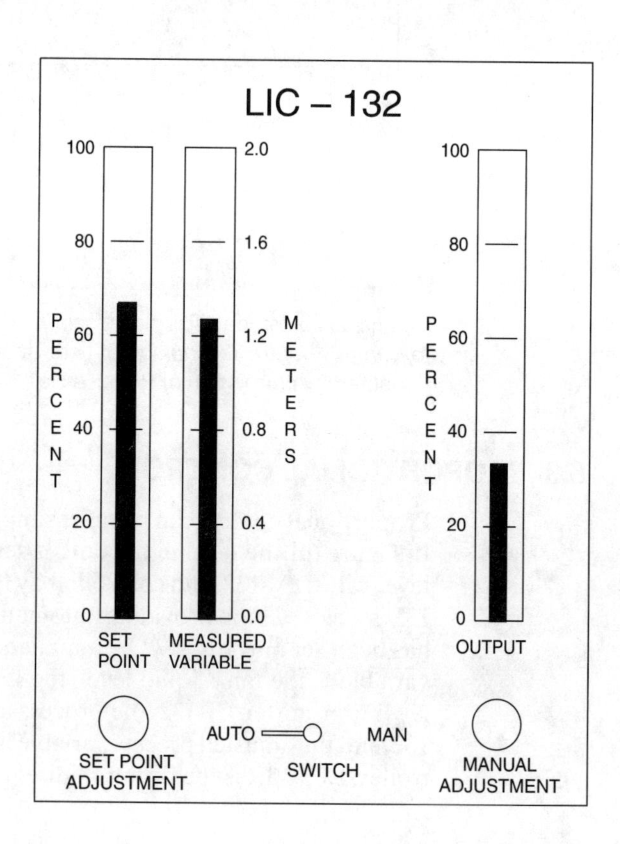

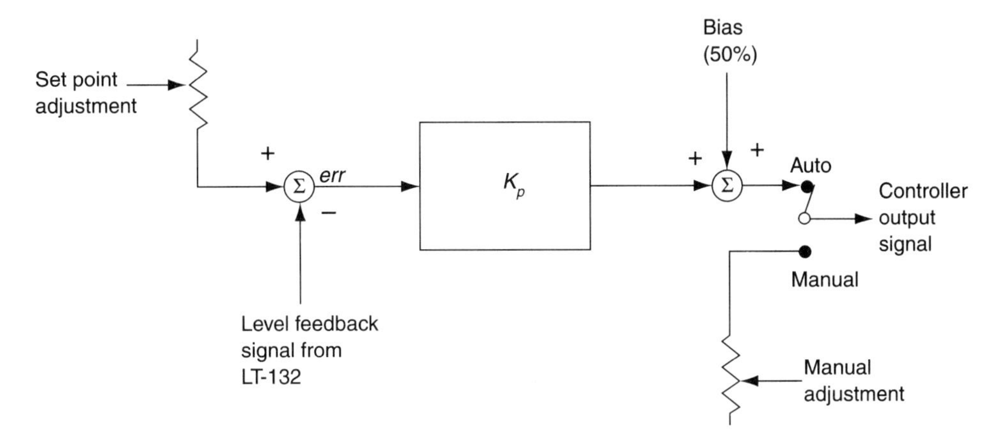

FIGURE 6.8
Block diagram of proportional level controller LIC-132

librium under these conditions for a long period of time. If the flow out of the tank is 59.6%, then the flow in must also be at 59.6% in order for the level to remain steady. For the controller to maintain the tank inflow valve at 59.6% open, then the controller output signal must be at 59.6%. The error must be at (59.6 – 50)% / 2.4 = 4%, and the feedback signal must be at (43 – 4)% = 39%. The tank level cannot be at the set point. An error must exist to have the loop and the tank remain in equilibrium.

With only the proportional control function, there is only one value of the set point and measured variable where the error is zero, and this value is the one that exists when the valve is held at the bias value (50%). Assume that the tank level is in stable equilibrium with the flow in just equal to the flow out. After a short time the operator upsets the control system by opening the manual valve on the outlet of the tank to 80% open. Assume that the flow out increases to 80% from 50%. Obviously, the tank level will start to drop. The feedback signal will fall below the set point and the error will increase. The controller output signal then will increase. Control valve LV-132 will open further, and, after awhile, the flow in will reach 80%. But the error must remain continually at 12.5% = (80 – 50) / 2.4, and the feedback signal must remain continually at 30.5 = 43 – 12.5, even though the level set point is at 43%. This continual existence of error, in this case 12.5%, is called *offset,* and it is a problem with proportional-only control. This problem of offset may be overcome by adding the integral function (discussed in section 6.4) to the proportional function. We also might whimsically consider having a little man inside the controller to adjust the bias so that it is slowly adjusted to produce the complete controller output signal, in this case 80% bias, and thus the error will remain at zero. Another possibility would be to have the operator reset the set point to 55.5% = 43 + 12.5, which would keep the level close to 43% when the flow out is close to 80%. However, whenever the flow out is changed, the set point would need to be reset in order to maintain the level at 43%. This may not be a problem if there is no concern for variations in level from about 23–63% due to flow out variations from 2–96%. Notice that, if we can maintain stable operation of the loop at very large values of gain, K_p greater than 50, there will only be very small variations in offset (error less than ±1%) as the flow out (and necessarily flow in) varies from 0–100%. However, the value of gain must be tuned to give a desirable response, and it often cannot be set to a large value because that will cause an undesirable response. The formula for the proportional only controller output is

$$CO = K_p\, e + bias$$

where: CO = controller output signal
e = error signal
$bias$ = bias signal

6.4 PROPORTIONAL PLUS INTEGRAL CONTROL

If we look closely at Figure 6.8 and imagine some improvement that will overcome the offset problem, we will focus on the error signal. We want this signal to be zero as much of the time as possible. That way the level will remain at the set point as much of the time as possible. If some automatic monitor watches the error signal, it could adjust the bias signal to maintain the error as close to zero as is possible. In what way should it adjust the bias signal if the error is not zero? If the error is zero, then the bias should simply hold whatever value it is at. The integral of the error with respect to time will provide both these features. If the error suddenly becomes some small positive value, then the integral action will start to increase the bias value, opening the valve, and raising the level slowly, and the error will then slowly decrease. When the error decreases to zero, the integral action will hold the bias at the value it has reached, and it will no longer change the bias. A negative value of error causes the integral action to decrease the bias in the same way. This addition to the proportional controller is shown in Figure 6.9. Figure 6.9 shows the proportional gain as K_p and the integral gain as K_I. These gains are tuned for the most desirable response as described in the subsection Desired Performance of a Control Loop (pp. 131–132). Notice that the integral action is affected by K_I and by K_p. The formula for the proportional plus integral controller output is

$$CO = K_p(e + K_I \int e\, dt)$$

where: CO = controller output
e = error signal

6.5 PROPORTIONAL PLUS INTEGRAL PLUS DERIVATIVE CONTROL

The addition of derivative action to the level controller improves its control performance in another way. The block diagram of the PID (proportional plus integral plus derivative) controller is shown in Figure 6.10. Derivative action anticipates overshoot

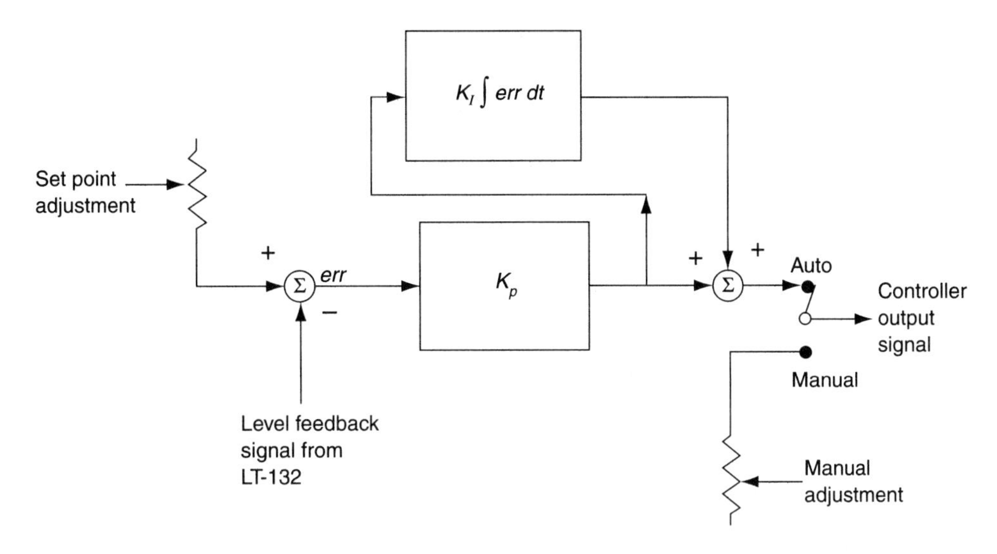

FIGURE 6.9
Block diagram of proportional plus integral level controller

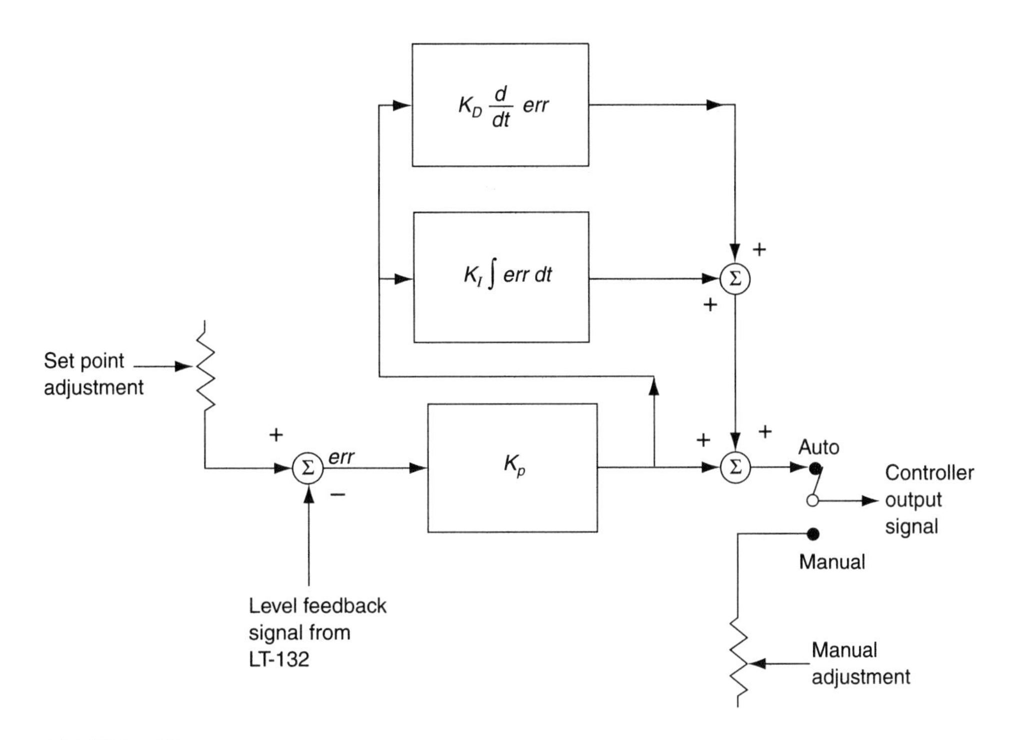

FIGURE 6.10
Block diagram of proportional plus integral plus derivative level controller

and tends to cut back the controller action as the measured variable rapidly approaches the set point. The rate of change (derivative with respect to time) of the error attempts to predict when the error will become zero and to compensate for expected overshoot. Notice that the value of K_p affects proportional, integral and derivative actions. The formula for the proportional plus integral plus derivative controller output is

$$CO = K_p(e + K_I \int e\,dt + K_D \frac{de}{dt})$$

6.6 LOOP TUNING

As mentioned under section 6.2, Loop Stability (pp. 130–135), there are several undesirable responses to loop disturbances. In order to achieve a desired response it is necessary to tune each loop. First, the desired response must be specified to the satisfaction of the operations personnel. Usually this desired response will fall into the types of response that are between critically damped and underdamped as shown in Figure 6.2. If the underdamped response has an open loop gain of 0.5 as shown in Figure 6.5, then the response will be called quarter amplitude damping with the amplitude of a positive-going half-wave having one-fourth the amplitude of the previous positive-going half-wave. So usually the desired response is specified as critically damped or quarter amplitude—or somewhere between these two responses. For this section of the chapter, assume the specified response that is desired is quarter amplitude response and that the tuning should result in this desired response to a disturbance.

Open Loop Gain Equals 1.0

As shown in Figure 6.4 the overall open loop gain—including the controller, control valve, process, and sensor—corresponding to a value of 1.0 occurs when oscillations of constant amplitude occur. So, if we adjust the loop gain, after the process and control loop have been constructed, by adjusting experimentally the controller gain in such a way that we discover the value of controller gain that provides oscillations of constant amplitude, then we can be fairly certain that we have established an overall open loop gain of 1.0. If the desired response is a one-quarter amplitude response requiring an open loop gain of 0.5, then it is only necessary to cut the controller gain to one-half of the value it had for constant amplitude oscillation.

In order to find the controller gain corresponding to an open loop gain of 1.0, it is necessary to ensure that the controller integral and derivative actions are turned off or at least set to have minimal effect. Then it is necessary to set the proportional gain to its minimum value. Under these conditions, when the control loop is closed (switched to auto control from manual control) and the set point is changed (either increased or decreased) by 10%, the response of the control loop should be similar to the overdamped response of Figure 6.2. Then the loop is switched to manual, the controller gain setting is doubled, the loop is switched back to auto, the set point is bumped 10% up or down, and the response is monitored. If the response is still overdamped or even underdamped with decreasing oscillations, then the procedure is repeated, again doubling the controller gain. This *ultimate cycle* procedure is continually repeated until constant oscillation occurs.* When constant oscillation is achieved it is necessary to record the controller gain setting as the ultimate gain, G_u, and the period of the oscillation as the ultimate period, P_u, in seconds per cycle.

Setting Proportional Gain

If the controller has proportional only control, then set the controller gain to $0.5 \times G_u$ in order to achieve quarter amplitude response. The overall closed loop gain becomes 0.5, which produces quarter amplitude response as shown in Figure 6.5.

Setting Proportional Plus Integral Gains If the controller has proportional and integral control, then set the proportional gain to $0.45 \times G_u$ and the integral gain to $1.2 / P_u$ repeats per second in order to achieve quarter amplitude response. The final value of offset should become zero.

Setting Proportional Plus Derivative Gains If the controller has proportional and derivative control, then set the proportional gain to $0.6 \times G_u$ and the derivative gain to $P_u / 8$ seconds in order to achieve quarter amplitude response. One of the advantages of derivative action is the higher gain of $0.6 \times G_u$, which provides faster response.

Setting Proportional Plus Integral Plus Derivative Gains If the controller has proportional and integral and derivative control, then set the proportional gain to $0.6 \times G_u$, the integral gain to $2.0 / P_u$ repeats per second, and the derivative gain to $P_u / 8$ seconds in order to achieve quarter amplitude response.

Further trial-and-error adjustment of the proportional gain only may be desirable to achieve the final desired response.

6.7 RATIO CONTROL

Ratio control is usually associated with blending two or more streams together in order to maintain a constant concentration of each chemical component in the overall blended fluid. It then follows an established recipe to achieve a final product. An example is maintaining the correct ratio of air flow to fuel flow into a furnace. The air ratio control loop senses the flow rate of fuel, which is independently varying in order to satisfy the demand for heat. The air ratio control loop then adjusts the flow rate of air to maintain the ratio of air flow in kg/sec to fuel flow in kg/sec at a constant value. Thus the flow rate of air follows the flow rate of fuel. The fuel flow is sometimes called the *wild flow,* since it appears to change in an independent way. The air ratio

---

*Material pertaining to the ultimate cycle procedure has been adapted for this section from R. Bateson. 1995. *Introduction to Control System Technology*. 5th ed. Upper Saddle River, NJ: Prentice Hall. Reprinted by permission of Prentice-Hall, Inc., Upper Saddle River, NJ.

control loop senses the air flow and the fuel flow and calculates the ratio of air flow to fuel flow; if this ratio is not at the ratio set point, then the air flow is adjusted to bring the ratio to the ratio set point. Another way of looking at it is to imagine that the set-point ratio is multiplied by the actual flow of the wild variable to obtain a flow rate set point for the controlled variable. For example, if the ratio set point of air flow in lb/hr to the fuel flow in lb/hr is 0.13, and if the actual flow of fuel is 1000 lb/hr, then the required flow of air is 130 lb/hr. Figure 6.11 shows the two flows and a ratio controller on the air flow rate. For ratio control there is instantaneous calculation of the air flow loop set point. The ratio controller can include proportional, integral, and derivative actions. Notice that FFIC stands for *flow fraction indicating controller,* which is the ratio controller.

6.8 CASCADE CONTROL

For a ratio control loop, the set point of the loop is made equal to the ratio setting times the measurement value of another loop. For cascade control, the set point of one loop, the slave loop, is made equal to the output from another loop controller, the master loop. The two loops are said to be *nested* (see Figure 6.12), with the slave process having an effect on the master process. An obvious cascade control system is the slave flow loop into a tank and the master level loop in the tank. For example, the flow control loop into the tank has a flow sensor, a flow controller, a flow control valve. The level control loop on the tank has a level sensor and a level controller. The output of the level controller adjusts the set point of the controller on the flow into the tank. The advantage of this arrangement is that any disturbances to the flow into the tank are instantly compensated for by the flow control loop. Whereas if the level control loop adjusted the inflow valve

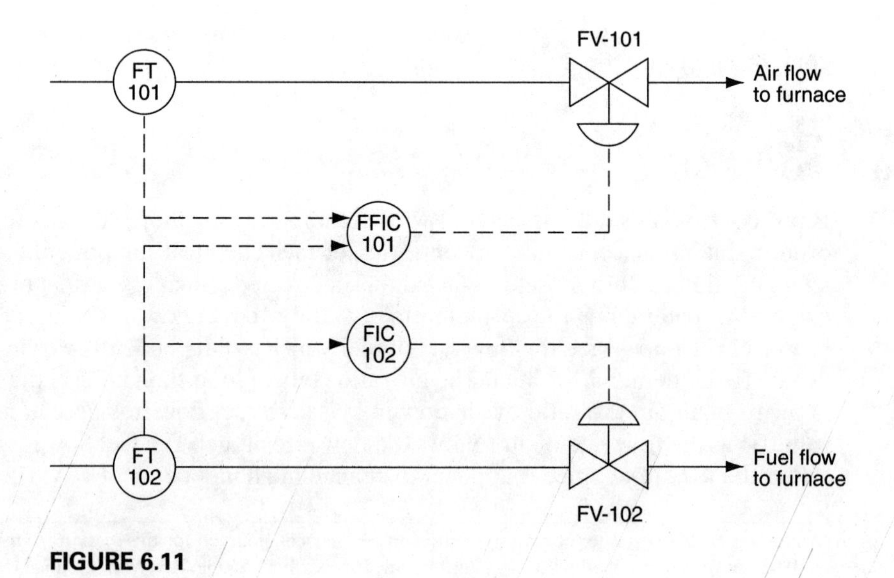

FIGURE 6.11
Ratio control

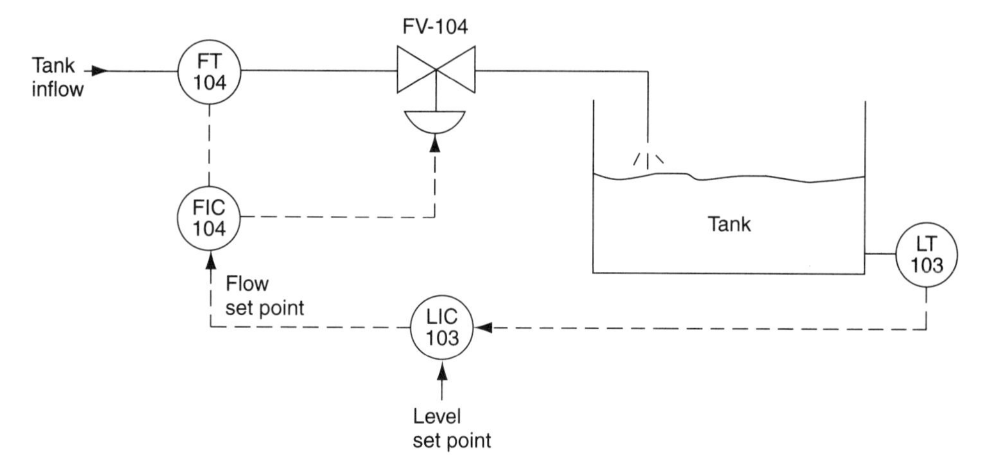

FIGURE 6.12
Cascade control

directly, then it may take a little while before enough change in the tank level occurred to cause the level control loop to compensate for the disturbances. Therefore cascade control is most useful for a very slow process, such as a level process, which also has a very fast process, that affects the slow process. Another example is the slow temperature process in a furnace and the fast fuel flow process. Disturbances to the fuel flow may not be detected by changes in the furnace temperature for a fairly long time, while a fuel flow control loop will compensate for them very quickly.

Generally, the master loop is the most important loop and it is desirable to keep its measured variable near its set point. The slave loop, however, is not so important. Therefore offset error in the slave loop is not of great concern. Thus integral action is usually not needed for the slave loop, which supports the view that the simpler the control system the better.

When tuning a cascade control system, it is necessary to tune the slave loop first. Then it becomes a correctly functioning component of the master loop when the master loop is tuned. This means that the master loop is put in manual (open loop) while the slave loop is tuned. Under this procedure each loop is tuned as described in section 6.5. The major design problem is associating the correct master and slave processes. This requires experience with the process.

PROBLEMS AND LAB ASSIGNMENTS

6.1 A proportional controller is shown in Figure 6.13. In the diagram shown in Figure 6.13:

r = set point in percent

c = meaured variable in percent

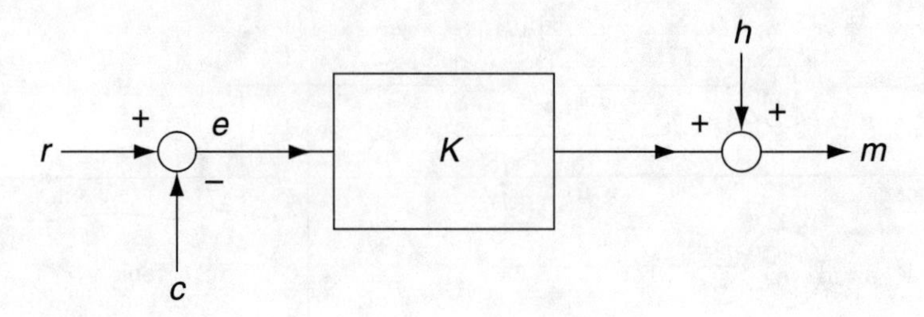

FIGURE 6.13
Diagram for problem 6.1

e = error in percent

K = controller gain

h = bias or manual reset in percent

m = manipulated variable in percent

Imagine that the control system has been functioning for the past 30 minutes at a steady value of $r = 60\%$, $m = 70\%$, $h = 50\%$, and $K = 2.0$. For this condition, find the values of e and c. If e is not 0%, why is it not?

6.2 A proportional-only controller has a proportional band setting of 20%. Proportional band is the inverse of gain. If proportional band = 100% / gain, then what is the gain of this controller? Draw the block diagram of this controller with its set point set at 15% and bias set at 50%. What is the offset error of this controller when its output is at 10%? What is the offset error when its output is at 50%? What is the offset error when its output is at 80%? How can we eliminate this offset error?

6.3 A tank level control loop is set up with a proportional controller as shown in Figure 6.14.

The flow control valve, LV-302, is at 80% open, and the flow-out valve is at 80% flow. The controller gain is set to $K = 5$ for stable operation, and the set point is set to 54%. What is the value of the measured variable signal from LT-302?

6.4 A tank level control loop is set up with a proportional controller as shown in Figure 6.14. The controller gain is set to $K = 4$ for stable operation, the set point is set to 33%, and the value of the measured variable from LT-302 is at 35.5%. What is the value of the signal that the controller is sending to the flow control valve LV-302?

6.5 For the controller of problem 6.1, what method is used to automatically obtain $e = 0\%$? Show this method by drawing a revised diagram similar to the one shown for problem 6.1 (Figure 6.13) and explain how it functions.

6.6 In Figure 6.15 the combined gain of the electropneumatic converter, FY- 101, the valve, FV-101, and the flow process, relates gallons per minute (GPM) of flow to the percent of controller output signal, and its value in this case is 3 GPM divided by percent. The vortex flow sensor gain, relating the percent of sensor output signal to flow in GPM, is 0.25% divided by GPM. The controller gain has not yet been

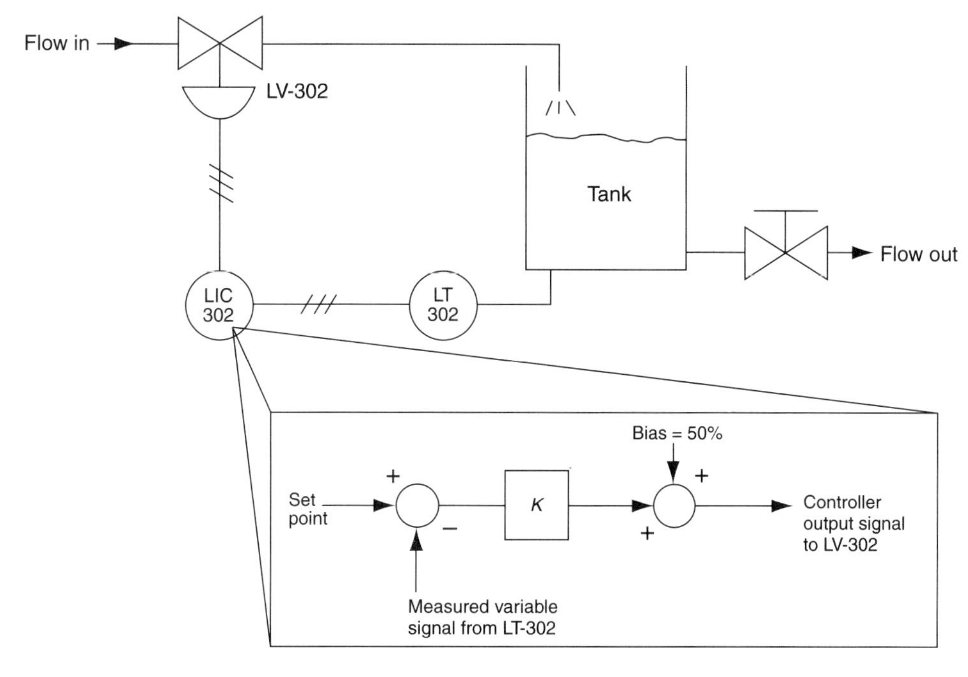

FIGURE 6.14
Diagram for problems 6.3 and 6.4

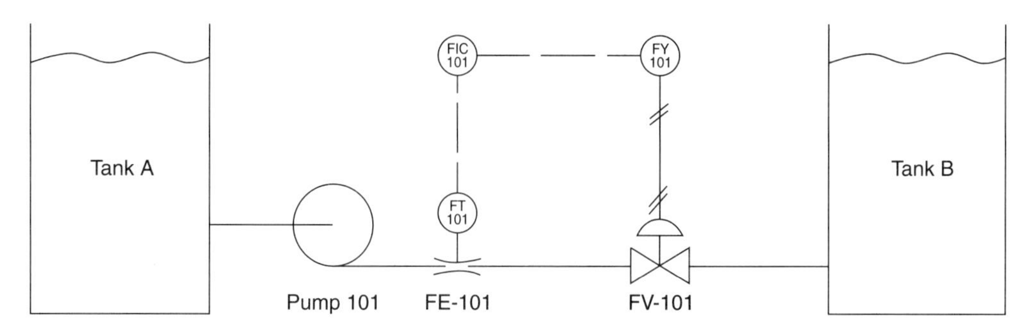

FIGURE 6.15
Diagram for problem 6.6

evaluated. Let us temporarily disconnect the flow sensor output signal from the feedback point of the controller to find the open loop gain. Draw the open loop block diagram with three blocks, one block for the sensor FT-101, one block for the process and the valve FV-101 and the electropneumatic converter FY-101, and one block for the controller. If we want the overall open loop gain from the set point signal to the sensor output signal to be 0.5 (% sensor output signal) / (% set point signal), then to what value do we set the controller gain?

6.7 A closed loop for flow control is being tuned to achieve quarter amplitude response. The loop includes a proportional-only controller, a flow control valve in a pipe to adjust the liquid flow rate, and a flow sensor in the pipe. Temporarily, the feedback signal from the flow sensor is disconnected from the controller and connected to an oscilloscope. A sine wave input is connected in place of the set point signal of the controller. The gain of the controller is set to 1.0. This gain remains constant at 1.0 with a phase shift of zero at all frequencies. The frequency of the sine wave at the set point is varied over a wide range. At the resonant frequency it is found that the phase shift around the open loop from the set point to the signal coming from the flow sensor lags by 180°. This means that the control valve, the process, and the sensor cause the phase to lag at that frequency by 180° from the phase of the sine wave entering the set point. At this frequency the amplitude of the signal from the flow sensor is twice the amplitude of the sine wave at the set point. Thus the open loop gain is 2 with a phase angle of –180°. If this loop is now closed by reconnecting the flow sensor signal to the controller, and the sine wave is disconnected from the set point, what response will the flow display to a manual step change in the set point? To what value should we adjust the controller proportional gain in order to achieve quarter amplitude response? In order to achieve critical damping, should we increase or decrease the gain of the controller from the setting for quarter amplitude response?

6.8 A blending system is required to blend three liquids. The normal recipe will be made up 50% by volume of liquid A, 30% by volume of liquid B, and 20% by volume of liquid C. Each liquid will be controlled by a control loop, including a magnetic flowmeter, a controller, and a control valve. The controllers for liquids B and C are ratio controllers. Their flows will be controlled automatically as a ratio of the flow of liquid A. The controller for liquid A will be a standard controller with operator adjustable set point. Draw a diagram, using ISA symbols of a tank, showing the three flows into the tank, and showing the three control loops with all the instruments associated with each loop. Show the interconnection of the loops to achieve ratio control. In a table, list the devices with the ISA symbols and numbers that you assign and a description of their function. Specify the fractional settings required for the two ratio controllers. To what fraction of the flow of liquid A is the flow of liquid B set in order to achieve the above recipe? To what fraction of the flow of liquid A is the flow of liquid C set? If the flow rate of liquid A is set to 25 lpm, what flow rate will be maintained for liquid B and liquid C?

6.9 A vessel with a jacket around it contains a liquid that flows in near the bottom and overflows to a pipe near the vessel's top. The vessel is heated by steam that flows into the jacket. The condensate in the jacket returns to the plant boiler. The temperature of the process liquid is sensed by an RTD located in the overflow pipe. This sensor is connected to a controller that adjusts the flow of steam into the jacket. This temperature control is less than satisfactory because of frequent fluctuations in the pressure of the steam. It appears as if better results would occur if the steam pressure could be adjusted to compensate for the variations. Draw a diagram using ISA symbols showing how an improved—but more complicated—control system could adjust quickly to compensate for these steam pressure disturbances. List each instrument device with its symbol and function. Explain how the control system will work, emphasizing the way improvements will reduce the effects of the steam pressure fluctuations.

6.10 Controller, Time Response, and Simulation

Objective: To simulate the time response of a controller to a varying error signal and compare it to the hand-calculated values.

Equipment: Spreadsheet program such as Corel Quattro Pro, Microsoft Excel, or Lotus 1-2-3 shown in Figures 6.16 and 6.17 with chart feature.

A	A	B	C	D	E	F	G	H	I	J	K	L	M	N	O	P
1	HUMBER	COLLEGE	- SCHOOL OF MANUFACTURING & DESIGN-					MANUAL FOR				0	0	35		0
2	COURSE:	CONTROL SYSTEMS					COURSE	NUMBER: ELEC 404				1	0	35		0
3	BY: ROY FRASER-		ELEC ENG TECHNY - CONTR			L SYSTEMS		DATE: DEC 1998				2	0	35		0
4												3	0	35		0
5	LAB 11 PID Controllers - Time Response - Simulation											4	0	35		0
6												5	0	35		0
7	Specify :	TLCGAIN,	TLCKI,	TLCDER,	VZERO,	ERROR	Signal in	P1..P100	for each second from 1 to 100			6	0.2	36.2		0.2
8	View :	time (col L), output signal (col N), and error (col P) on graph at D90 when macro is run from D17.										7	0.4	37.4		0.4
9		Ctrl>Break to stop after one pass when JC = 100 secs.										8	0.6	38.6		0.6
10												9	0.8	39.8		0.8
11												10	1	41		1
12												11	1.5	44		1.5
13			LASTIM	36161.72	Fraction of Day, prev time through Main							12	2	47		2
14			SECS	62546	seconds of day							13	2.5	50		2.5
15			DT	1	step size, secs							14	3	53		3
16												15	3.5	56		3.5
17			\Simstrt	{;skip}								16	4	59		4
18				{\Init}					Initialize			17	4.2	60.2		4.2
19				{;skip}								18	4.4	61.4		4.4
20			SimWait	{If @NOW<LASTIM+.00001157407}{Branch SimWait}								19	4.6	62.6		4.6
21				{Let LASTIM,@NOW}								20	4.8	63.8		4.8
22				{;skip}								21	5	65		5
23				{WindowsOn}								22	5.1	65.6		5.1
24				{calc}					show graph update			23	5.2	66.2		5.2
25				{;skip}					Update			24	5.3	66.8		5.3
26				{Let SECS,(@NOW-@INT(@NOW))*86400}								25	5.4	67.4		5.4
27				{;skip}								26	5.5	68		5.5
28				{;skip}								27	5.6	68.6		5.6
29				{;skip}								28	5.7	69.2		5.7
30				{\LIC}					Update Controller			29	5.8	69.8		5.8
31				{;skip}								30	5.9	70.4		5.9
32				{;skip}								31	6	71		6
33				{;skip}								32	6.1	71.6		6.1
34				{;skip}								33	6.2	72.2		6.2
35				{\Graph}								34	6.3	72.8		6.3
36				{;skip}								35	6.4	73.4		6.4
37				{Branch SimWait}								36	6.5	74		6.5
38												37	6.6	74.6		6.6
39												38	6.7	75.2		6.7
40				INITIALIZE								39	6.8	75.8		6.8
41												40	6.9	76.4		6.9
42			\Init	{;skip}								41	7	77		7
43				{;skip}								42	7	77		7
44				{;skip}								43	7	77		7
45				{;skip}								44	7	77		7
46				{;skip}								45	7	77		7
47				{;skip}								46	7	77		7
48				{;skip}								47	7	77		7
49				{;skip}								48	7	77		7
50				{Let LASTIM,@NOW}								49	7	77		7
51				{;skip}								50	7	77		7
52				{;skip}								51	6.5	74		6.5
53				{;skip}								52	6	71		6
54				{;skip}								53	5.5	68		5.5
55				{Let TLCERS,0}								54	5	65		5
56				{Let TLCSET, TLTSIG}					Init Controller			55	4.5	62		4.5
57				{Let JC,0}								56	4	59		4
58				{Let MC,0}								57	3.5	56		3.5
59				{;skip}								58	3	53		3
60				{Return}								59	2.5	50		2.5

FIGURE 6.16
Quattro Pro versions 6 to 9 macro (continued on Figure 6.17)

A	A	B	C	D	E	F	G	H	I	J	K	L	M	N	O	P
61												60	2	47		2
62												61	1.5	44		1.5
63					CONTROLLER							62	1	41		1
64			TLTSIG	50								63	0.5	38		0.5
65			TLCSET	50	set point, %							64	0	35		0
66			TLCGAIN	6	gain							65	-0.2	33.8		-0.2
67			TLCKI	0	integral action, repeats/sec							66	-0.4	32.6		-0.4
68			TLCDER	0	derivative action, seconds							67	-0.6	31.4		-0.6
69			TLCPER	0	present error, %							68	-0.8	30.2		-0.8
70			TLCLER	-10	last error, %							69	-1	29		-1
71			TLCOS	-25	controller output signal, %							70	-1.2	27.8		-1.2
72			TLCERS	201	sum of errors							71	-1.4	26.6		-1.4
73			TLCINE	0	integral effect							72	-1.6	25.4		-1.6
74			TLCDRE	0	derivative effect							73	-1.8	24.2		-1.8
75			VZERO	35								74	-2	23		-2
76			\LIC	{;skip}								75	-2	23		-2
77				{Let TLCERS,TLCERS+TLCPER}								76	-2	23		-2
78				{Let TLCINE,TLCKI*DT*TLCERS}								77	-2	23		-2
79				{Let TLCDRE,TLCDER*(TLCPER-TLCLER)/DT}								78	-2	23		-2
80				{Let TLCOS,+VZERO+TLCGAIN*(TLCPER+TLCINE+TLCDRE)}								79	-2	23		-2
81				{Let TLCLER,TLCPER}								80	-2	23		-2
82				{Return}								81	-2	23		-2
83			MC	0								82	-2	23		-2
84			JC	1								83	-2	23		-2
85			\Graph	{WindowsOff}								84	-2	23		-2
86				{Put M1..M102,0,+JC,+TLCPER}								85	-2	23		-2
87				{Put N1..N102,0,+JC,+TLCOS}								86	-2	23		-2
88				{Let JC,JC+1}								87	-2	23		-2
89				{If JC=9}{Let MC,1}{Let MC,0}								88	-2	23		-2
90				{If JC>100}{Let JC,1}								89	-2	23		-2
91				{;skip}								90	-2	23		-2
92				{Return}								91	-2	23		-2
93												92	-2	23		-2
94												93	-2	23		-2
95												94	-2	23		-2
96												95	-2	23		-2
97												96	-2	23		-2
98												97	-2	23		-2
99												98	-2	23		-2
100												99	-2	23		-2
101												100	-10	-25		-10
102												101	-10	19.4		-10
103												102	-10	19		-10
104												103	-10	18.6		-10
105												104	-10	18.2		-10
106												105				-10
107																-10
108																-10
109																
110																
111																
112																
113																
114																
115																
116																
117																
118																
119																
120																

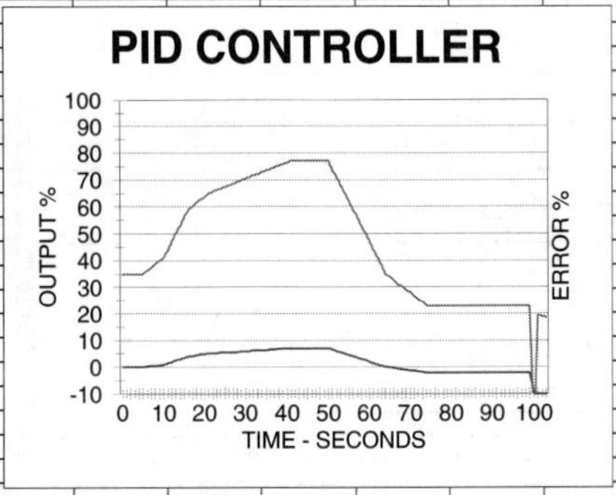

FIGURE 6.17
Quattro Pro macro (continued from Figure 6.16)

Procedure: First create the spreadsheet (with labels in column C) shown in Figures 6.16 and 6.17, including column L. Key into column P of the spreadsheet the error value for each second for a period of 100 seconds. Key into TLCGAIN the controller gain, into TLCKI the controller integral rate in repeats per second, into TLCDER the controller derivative time in seconds, into VZERO the controller bias in percent. Watch the resulting controller output on the graph and in column N after you play the macro from cell D17.

The chart or graph is made by selecting columns L (time in seconds), M (error curve), and N (output curve) from 1 to 101 on the spreadsheet. Then create a line or area type chart. Next click on regular line type chart. Choose the colors for the lines or default to the computer's choice. Next put in the main title and X and Y axis titles. Finish by positioning the chart on the spreadsheet where you want the top left corner to be.

For an error signal varying linearly between each of the following points:

secs:	0	5	10	16	21	41	50	64	74	99
err (%):	0	0	1	4	5	7	7	0	–2	–2

Find the output responses in percent of the following controllers:

P (gain)	**I (rep/sec)**	**D (secs)**	**Bias (%)**
4	0	0	50
4	0.01	0	50
4	0	10	50
4	0.01	10	50

Check by hand that they are correct at 40 seconds.

Submit printouts of the graphs with your hand calculations in your report.

REFERENCES

R. Bateson. 1996. *Introduction to Control System Technology.* 5th ed. Upper Saddle River, NJ: Prentice Hall.

Appendix A

ISA SYMBOLS AND TEMPLATES FOR CALIBRATION CERTIFICATES AND GRAPHS*

The following are featured in this appendix:

- Origin, purpose, and scope of ISA symbols
- Identification format and table of identification letters for control loop functions and devices
- Typical instrument tag numbers
- Instrument signal line symbols
- Typical instrument symbols on drawings
- Blank instrument calibration certificate
- Blank calibration graph, including deviation graph

A.1 ORIGIN, PURPOSE, AND SCOPE

The concepts and their symbols presented in the ISA-S5.1 standard are used extensively by industry to designate instrumentation and control systems in a standard format on engineering drawings. The heavy chemical processing industries, including petroleum refineries, electric power generating plants, paper mills, and mineral refineries, have used these concepts and symbols for more than 40 years, and many newer industries have adopted them.

The major unit common to all instrumentation is called a *loop*. A complete control loop includes instrumentation for measurement and control of a process

This appendix is adapted from ANSI/ISA-S5.1-1984, *Instrumentation Symbols and Identification.* Reprinted by permission.

variable. It is called a *control loop* because the signal corresponding to the measurement is transmitted to the controller from the process-variable measuring instrument. Then the controller interprets that signal and generates a corresponding control signal that it transmits to the process-variable adjusting device. Then the process variable adjusts to the desired value of the controller, and the process variable measuring instrument transmits the revised signal back to the controller. In this way, the signal travels around the control loop. The complete loop includes the process, the process measuring device, the controller, and the process adjusting device. The process is generally considered separate from the instrumentation that forms the other three parts of the loop.

There are two kinds of loops: the complete control loop and the semi-loop. A semi-loop includes the process and only the instrumentation for the measurement of the process variable. A semi-loop has no automatic control function.

A.2 THE IDENTIFICATION FORMAT

In large process plants there are usually many processes, and each process has many loops, and each loop has several functions or blocks that make up the loop. In order to describe each loop on the drawings, the function of each block in the loop must be identified separately. This standard describes the accepted designation for the function of each block in the loop and the accepted method of distinguishing one loop from another. Usually the process designers, a group separate from the instrumentation designers, have identified the process equipment with a distinct set of letters and numbers. For this reason the complete designation of each loop function includes a suffix for the process equipment identification, followed by the loop and the block identification. This complete designation is called the *expanded tag number,* and it includes the suffix for the process equipment identification. The tag number itself does not include the suffix for the process equipment identification.

The first letter of the tag number of a loop is selected to correspond to the measured process variable, such as *T* for temperature, *F* for flow, or *P* for pressure, per Table A.1. Each tag number designating the function of each block in that loop commences with that same first letter. Thus the loop or semi-loop is identified primarily by its measurement.

The succeeding letters in the tag number describe the function or functions of that block of the loop, such as *T* for transmit, *I* for indicate, *R* for record, or *C* for control.

The sequence of the tag number begins with a first letter for the measurement according to Table A.1, followed by succeeding letters in any order for the function or functions of the block also according to Table A.1. The modifier letter immediately follows the letter that it modifies, and the control valve function is designated *CV* or sometimes just *V.*

The loop is identified by the first letter and a number that follows the functional letters. Therefore, parallel numbering, such as TIC-100, FRC-100, and LT-100, distinguishes one loop from another. Serial numbering, however, such as TIC-100, FRC-101, and LT-102, provides a distinct number for each loop and is equally acceptable.

TABLE A.1
Instrumentation identification letters

First Letter		Succeeding Letters		
Measured or Initiating Variable	Modifier	Readout or Passive Function	Output Function	Modifier
A—Analysis		Alarm		
B—Burner Flame				
C—Conductivity (Electrical)			Control	
D—Density	Differential			
E—Voltage		Primary element		
F—Flow Rate	Ratio or fraction			
G—Gauging		Glass		
H—Hand				High
I—Current (electrical)		Indicate		
J—Power	Scan			
K—Time or schedule	Time rate of change		Control station	
L—Level		Lamp		Low
M—Moisture	Momentary			Middle
N				
O	Orifice			
P—Pressure		Point (test)		
Q—Quantity	Totalize or integrate			
R—Radiation		Record or print		
S—Speed or frequency	Safety		Switch	
T—Temperature			Transmit	
U—Multivariable				
V—Vibration or viscosity			Valve	
W—Weight or force		Well		
X—Unclassified or X axis				
Y—Event or Y axis			Relay or compute	
Z—Position			Drive	

The user of this book may choose to apply blank positions in Table A.1 for his or her own use if that use does not fit one of the positions that are listed in the table.

A.3 TYPICAL TAG NUMBER*

A typical tag number is

TIC-103 Temperature Indicate Control (loop number) 103,

*The discussion and excerpts from the ISA-S5.1 standard in this appendix provide only a brief introduction to this widely used standard. For more information, or to obtain a copy, contact ISA, 67 Alexander Drive, Research Triangle Park, NC 27709, USA, telephone (919) 549-8411, or www.isa.org.

where TXX-103 identifies the loop by first letter and number,
and IC in place of XX identifies the functions by succeeding letter.

A typical expanded tag number is

T107-TIC-103

where T107 identifies process Tank number 107.

Note: Hyphens are optional.

A.4 INSTRUMENT LINE SYMBOLS

Figure A.1 shows instrument line symbols. All lines are to be fine in relation to process piping.

A.5 TYPICAL USE OF INSTRUMENT SYMBOLS ON DRAWINGS

Figures A.2 and A.3 show the typical use of instrument symbols on drawings.

A.6 INSTRUMENT CALIBRATION CERTIFICATE AND GRAPH

Figure A.4 shows a blank instrument calibration certificate. A blank calibration graph with deviation graph is shown in Figure A.5.

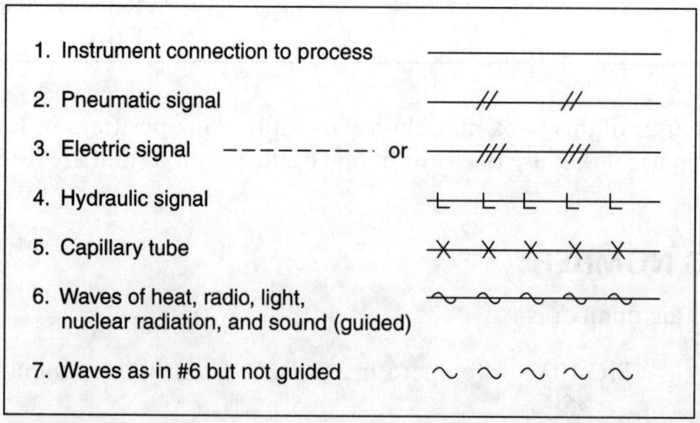

1. Instrument connection to process
2. Pneumatic signal
3. Electric signal or
4. Hydraulic signal
5. Capillary tube
6. Waves of heat, radio, light, nuclear radiation, and sound (guided)
7. Waves as in #6 but not guided

FIGURE A.1
Instrument line symbols

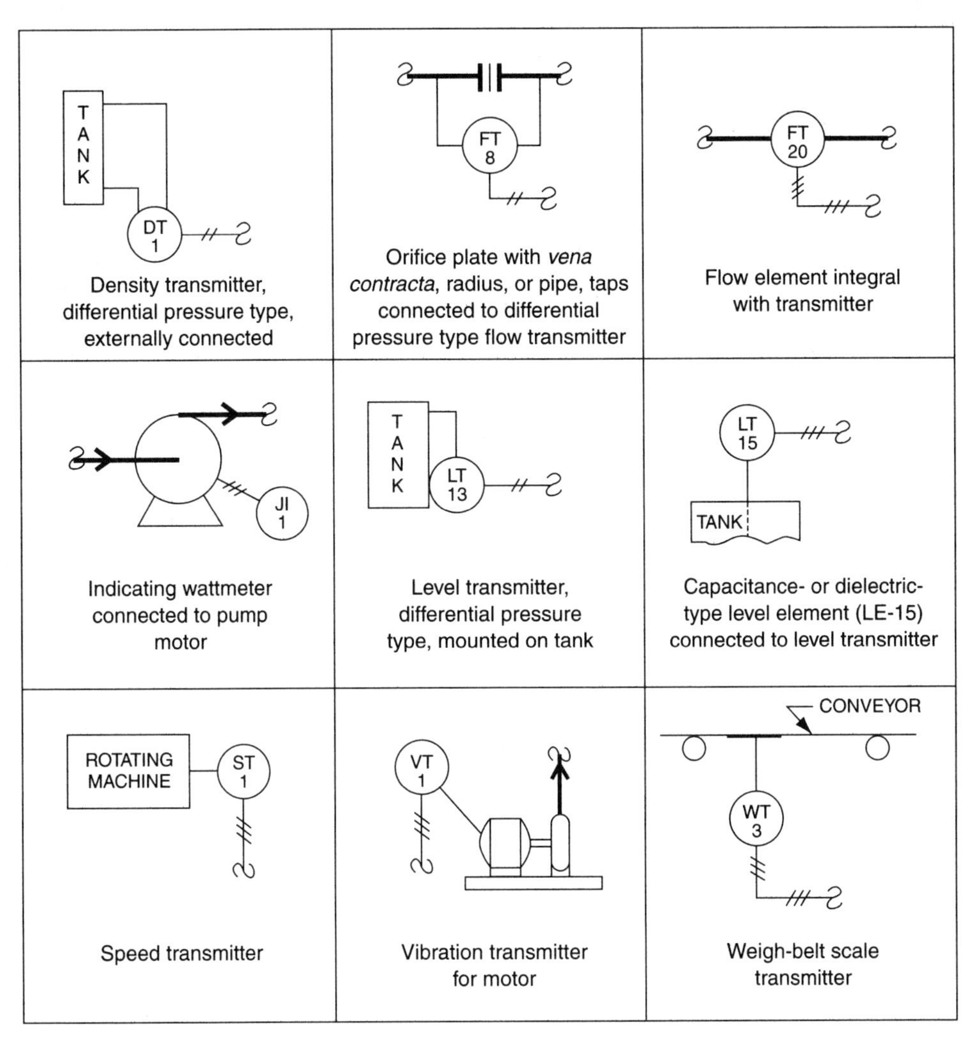

FIGURE A.2
Instrument symbols on drawings *(continued on Figure A.3)*

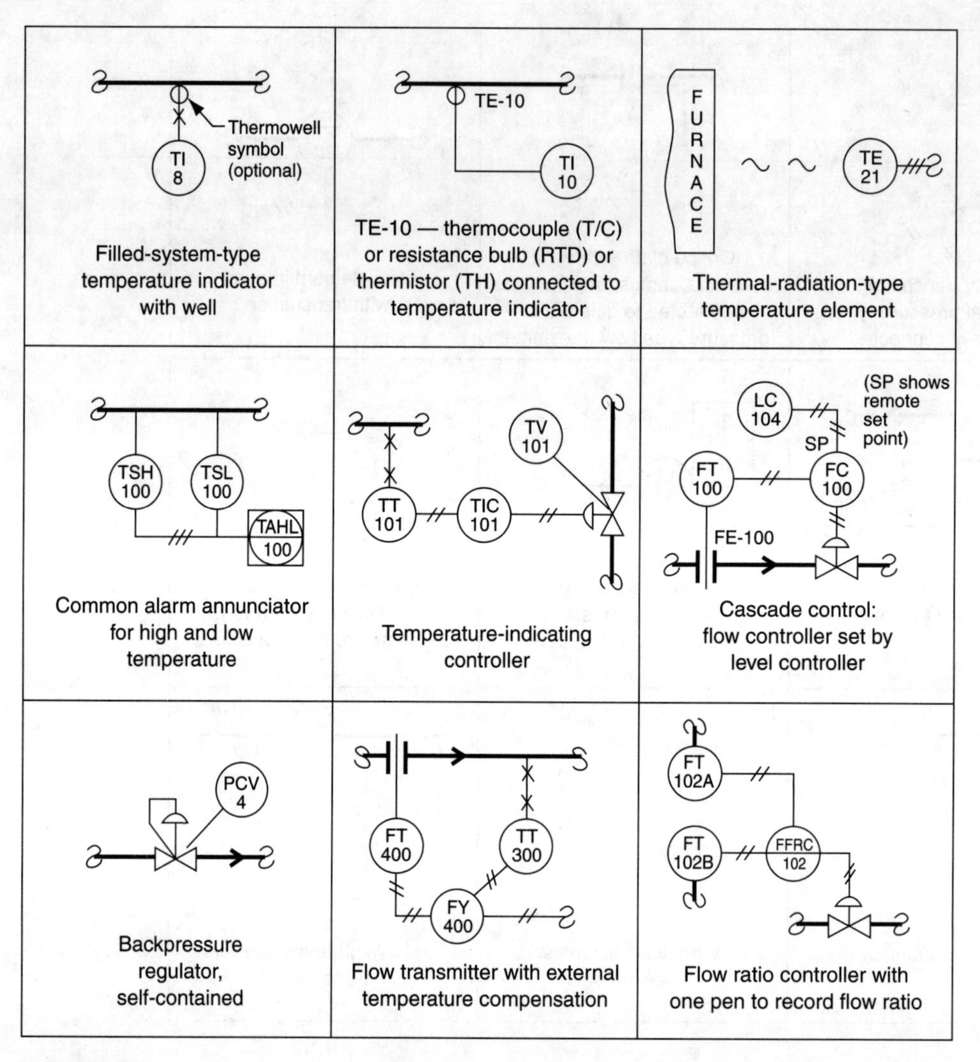

FIGURE A.3
Instrument symbols on drawings *(continued from Figure A.2)*

INSTRUCTMENT CALIBRATION CERTIFICATE						
Identification:				Input Range:		
				Output Range:		
Date:		Mfr:		Model:		
A Standard Input Signal	B Percent Input Signal	C Desired Output Signal	D Actual Output Signal	E Actual Output Error	F Percent Output Error	G Percent Output Signal
Remarks:						
				Signature:		

FIGURE A.4
Instrument calibration certificate

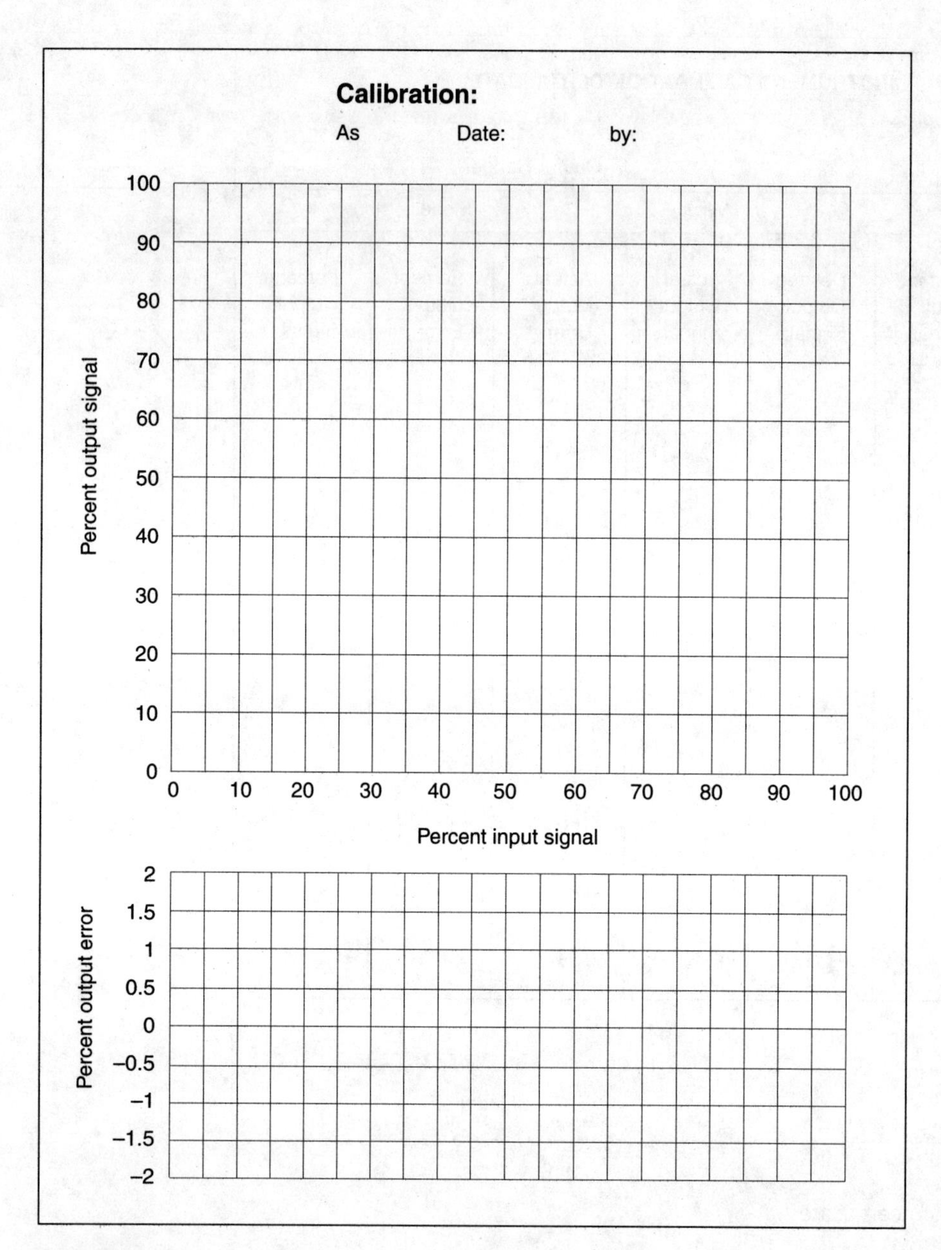

FIGURE A.5
Calibration and deviation graphs

Appendix B

CONTROL VALVE BODIES, ACTUATORS, POSITIONERS, AND ACCESSORIES

See Figures B.1–B.9 (starting on page 160).

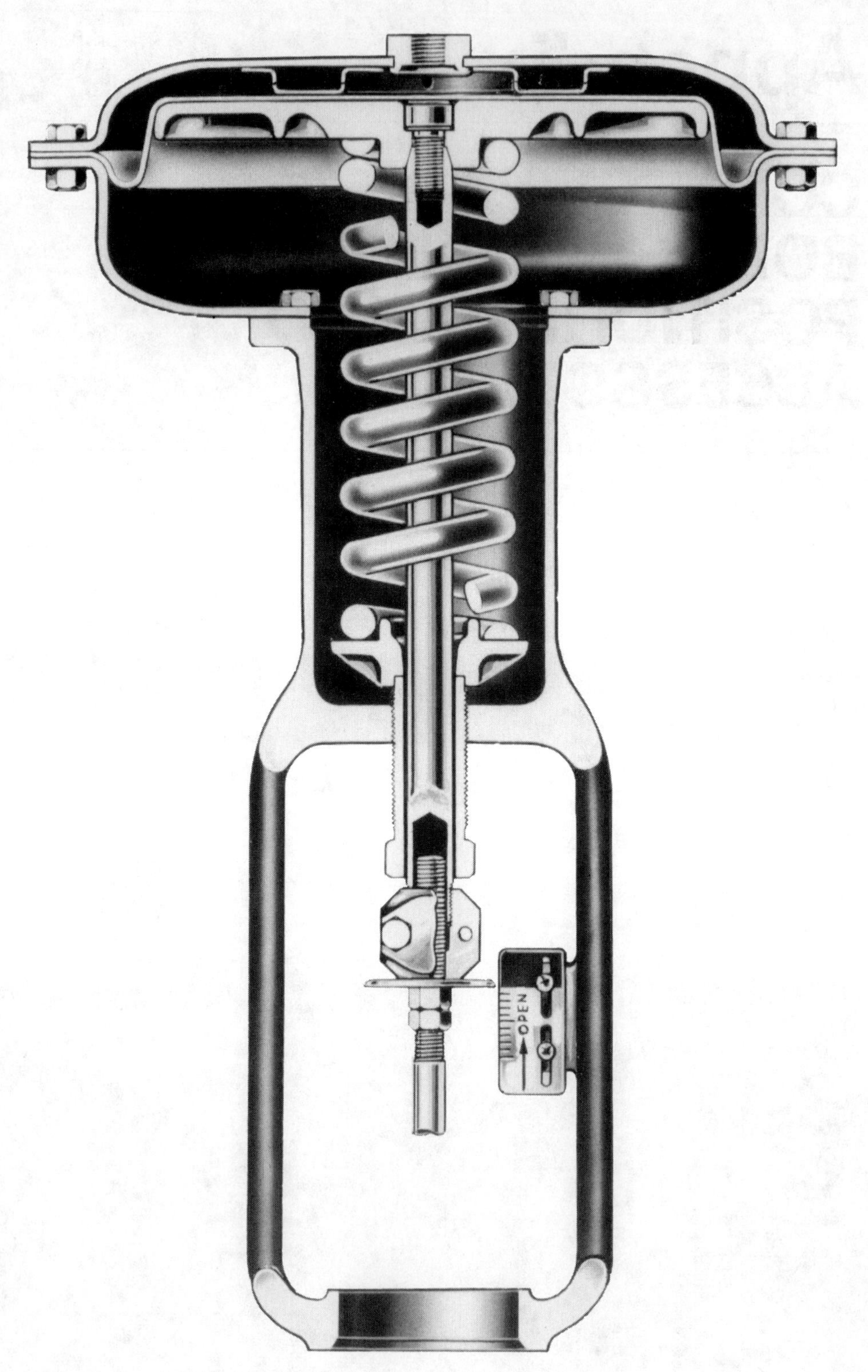

FIGURE B.1
Pneumatic diaphragm actuator *(Courtesy of Fisher Controls International, Inc.)*

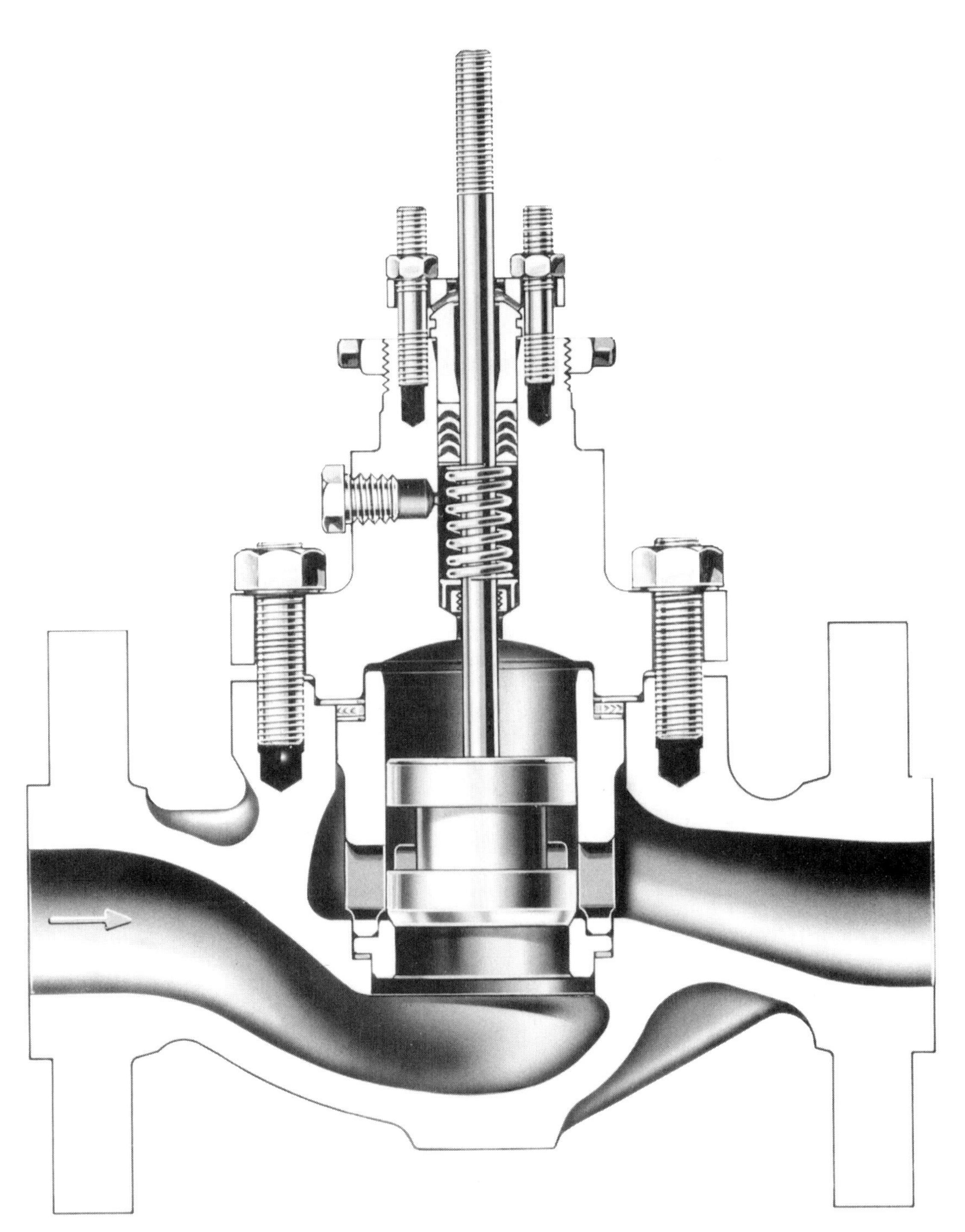

FIGURE B.2
Cage control valve body assembly (*Courtesy of Fisher Controls International, Inc.*)

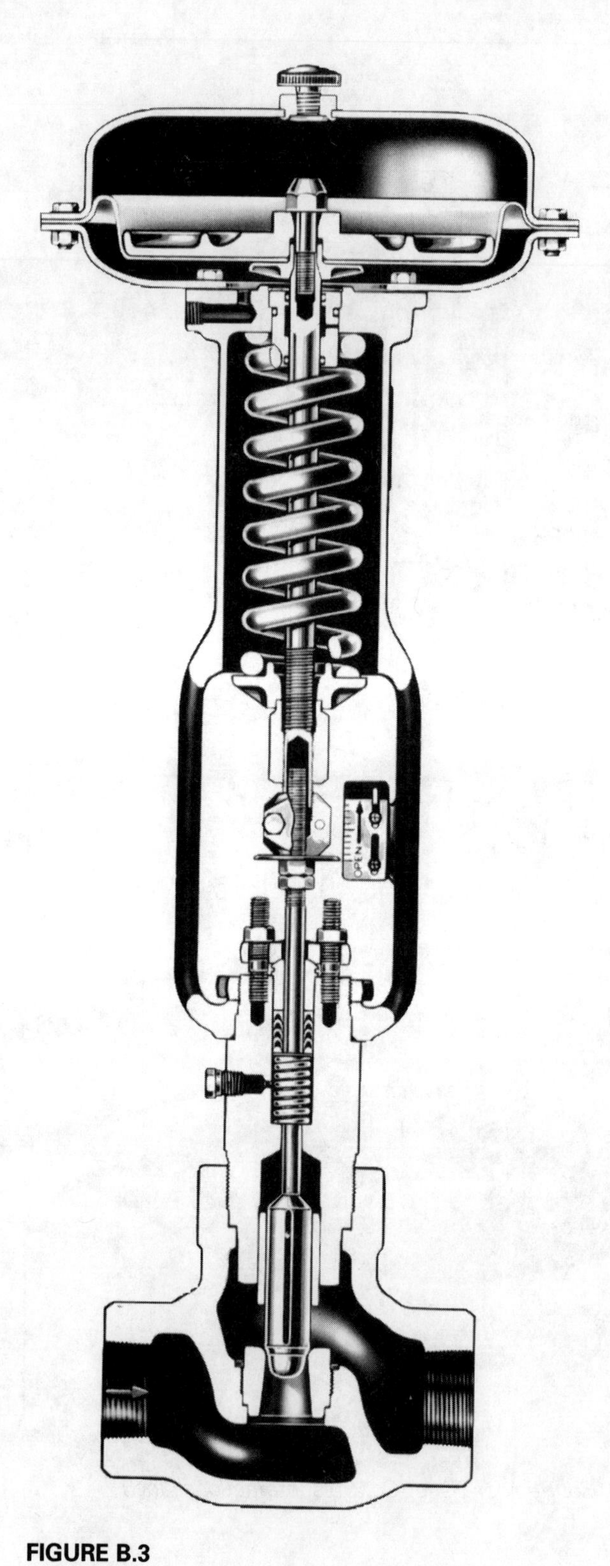

FIGURE B.3
Adjustable trim size, globe-style control valve with diaphragm *(Courtesy of Fisher Controls International, Inc.)*

FIGURE B.4
Butterfly control valve with rotary-shaft pneumatic piston actuator *(Courtesy of Fisher Controls International, Inc.)*

FIGURE B.5
Ball control valve with rotary-shaft pneumatic piston actuator *(Courtesy of Fisher Controls International, Inc.)*

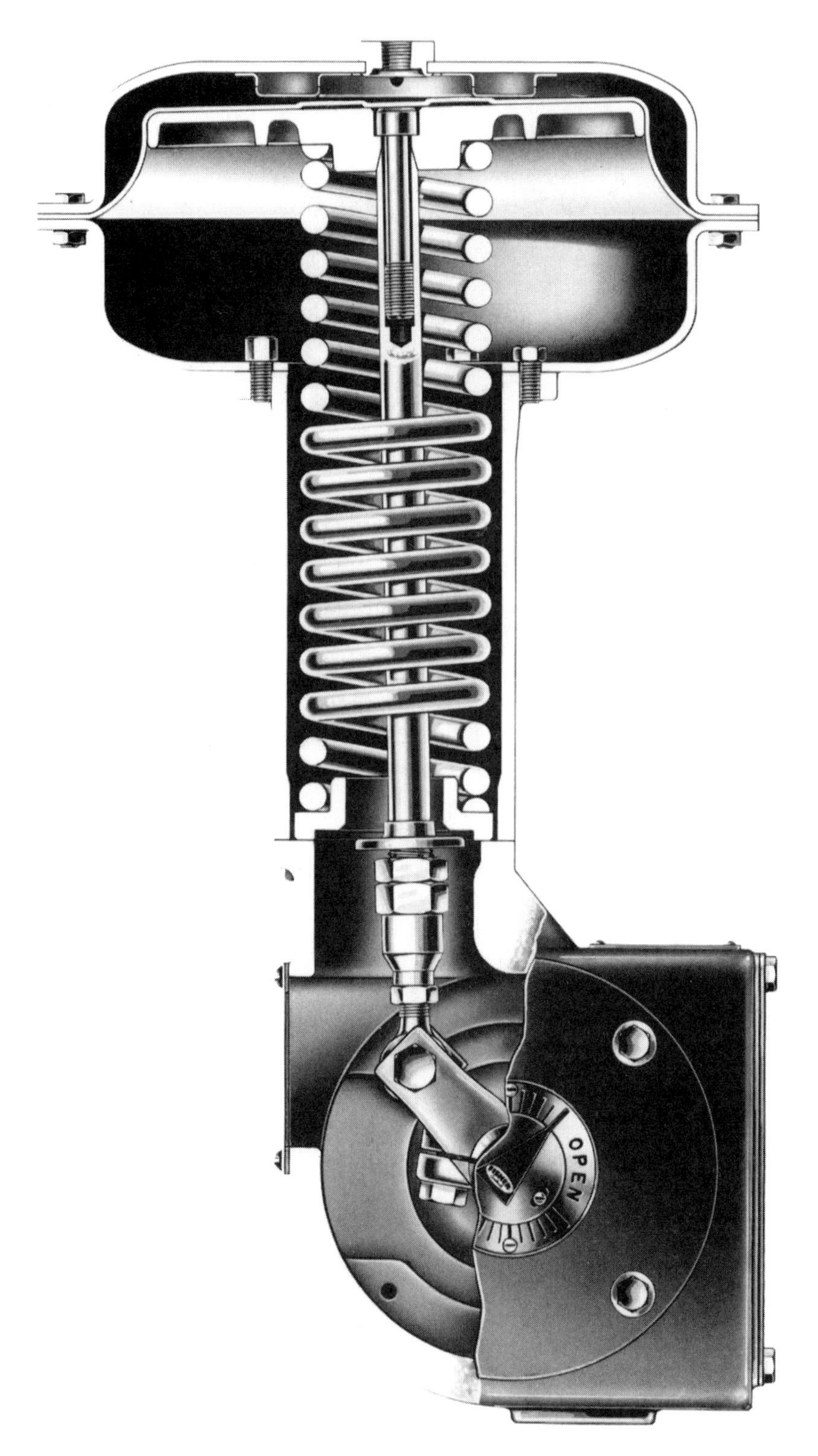

FIGURE B.6
Pneumatic diaphragm rotary actuator *(Courtesy Fisher Controls International, Inc.)*

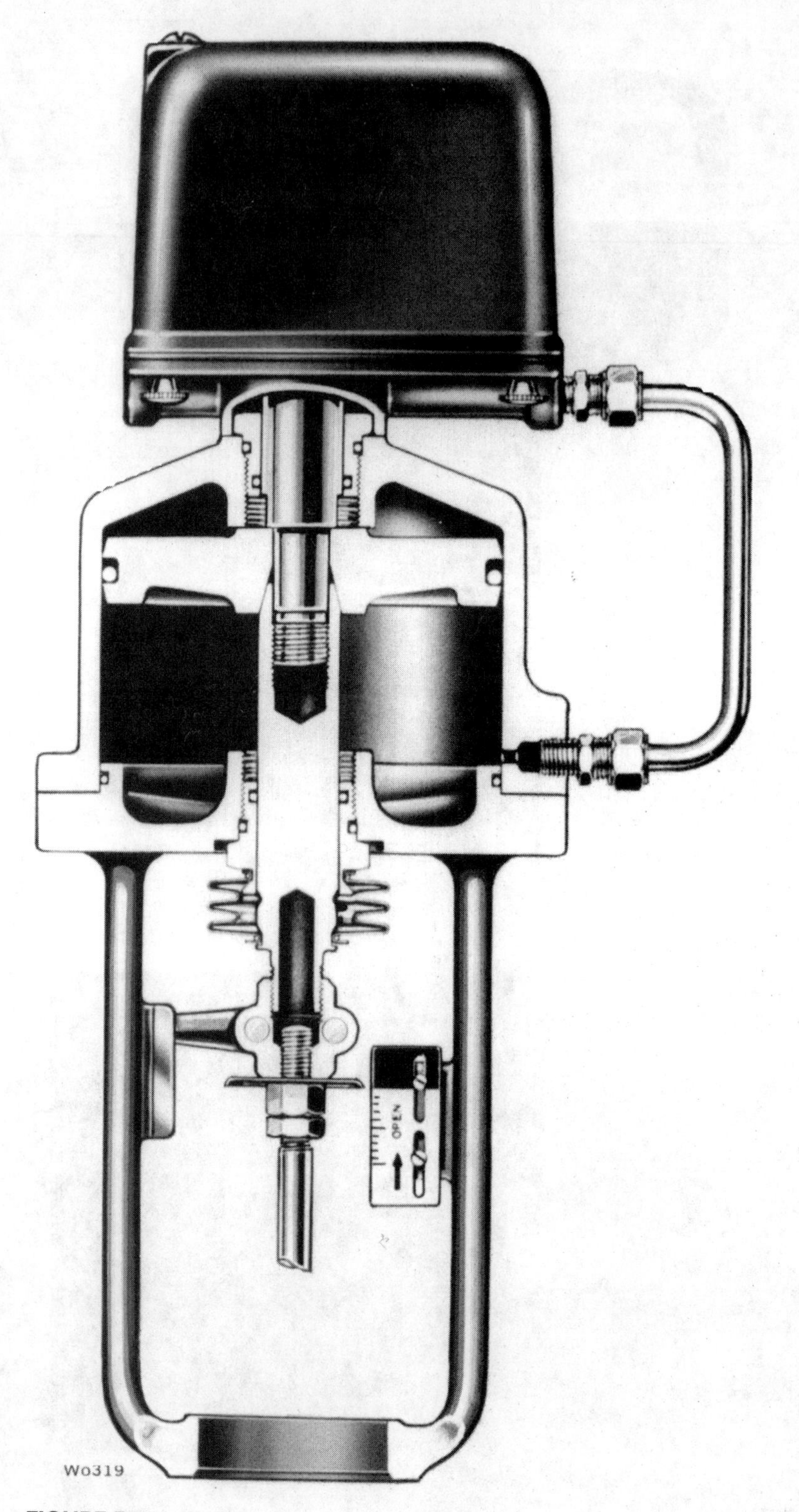

FIGURE B.7
Double-acting pneumatic piston actuator *(Courtesy of Fisher Controls International, Inc.)*

FIGURE B.8
Control valve diaphragm actuator with pneumatic positioner *(Courtesy Fisher Controls International, Inc.)*

FIGURE B.9
Pneumatic diaphragm actuator with electropneumatic positioner *(Courtesy of Fisher Controls International, Inc.)*

Appendix C

ANSWERS TO SELECTED PROBLEMS

CHAPTER 1—INTRODUCTION

Problem 1.1

Annual Losses	
Before new system	$1,260,000
After new system	–225,000
Reduction of Losses	1,035,000
Minus maintenance for new system	–220,000
Total annual savings of new system	815,000

One Time Only Costs	
New system	$2,875,000
Spare parts	260,000
Installation	470,000
Training	+185,000
Total new system initial costs	3,790,000

$$\text{Annual return on investment} = \frac{815,000 \times 100\%}{3,790,000} = 21.5\%$$

$$\text{Payback period} = \frac{\$3,790,000}{\$815,000/\text{year}} = 4.65 \text{ years}$$

Problem 1.4

The P&ID for temperature loop T-165 is shown in Figure C.1.

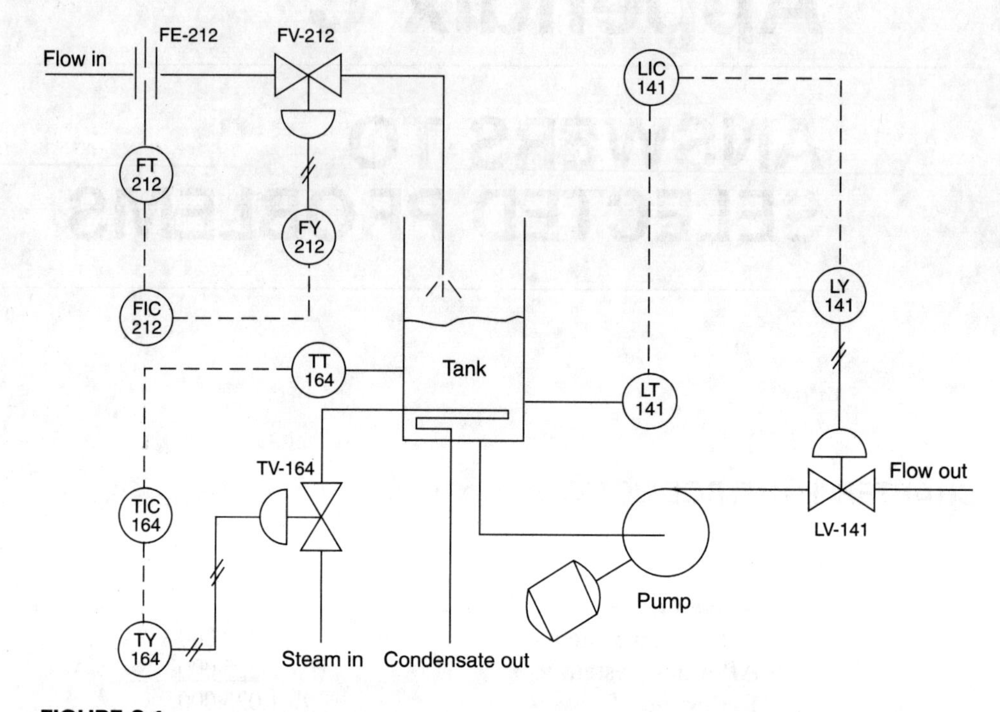

FIGURE C.1
Temperature Loop T-164 added to Figure 1.3

CHAPTER 2—MEASURING SENSORS

Problem 2.1

"As-found" instrument calibration certificate is shown in Figure C.2 with graphs in Figure C.3.

Problem 2.4

310°C gives 16.879 mv, 315°C gives 17.155 mv, and with the reference junction at 0°C, the data acquisition system showing 16.93 mv, the hot junction is at:

$$310°C + (315 - 310) \times (16.93 - 16.879) / (17.155 - 16.879) = 310.924°C$$

Problem 2.6

180°C produces 9.667 mv with the reference junction at 0°C. If the reference junction is at 15°C producing 0.762 mv, then the J (iron/constantan) thermocouple will provide 9.667 – 0.762 = 8.905 mv to the data acquisition system.

INSTRUMENT CALIBRATION CERTIFICATE

Identification: T14C-PT-325 | Input Range: 0–50 psig | Output Range: 4–20 ma

Date: 21 Sept 98 | Mfr: Foxboro | Model: IGP10-D20D1C

A Standard Input Signal (psig)	B Percent Input Signal (%)	C Desired Output Signal (ma DC)	D Actual Output Signal (ma DC)	E Actual Output Error (ma DC)	F Percent Output Error (%)	G Percent Output Signal (%)
1.008	2.016	4.323	4.233	–0.09	–0.5625	1.456
13.108	26.216	8.195	7.907	–0.288	–1.8	24.419
26.520	53.040	12.486	12.197	–0.289	–1.806	51.231
37.175	74.350	15.896	15.572	–0.324	–2.025	72.325
50.000	100.00	20.000	20.111	0.111	0.694	100.694

Remarks: As Found

Signature: Roy Fraser

FIGURE C.2
Completed certificate for Problem 2.1

Problem 2.8

RTD resistance = 327.9 mv / 1.5 ma = 218.6 Ω. 315°C has 217.36 Ω, 320°C has 219.13 Ω. 218.6 Ω is produced by

$$315°\text{C} + (320 - 315) \times (218.6 - 217.36) / (219.13 - 217.36) = 318.50°\text{C}$$

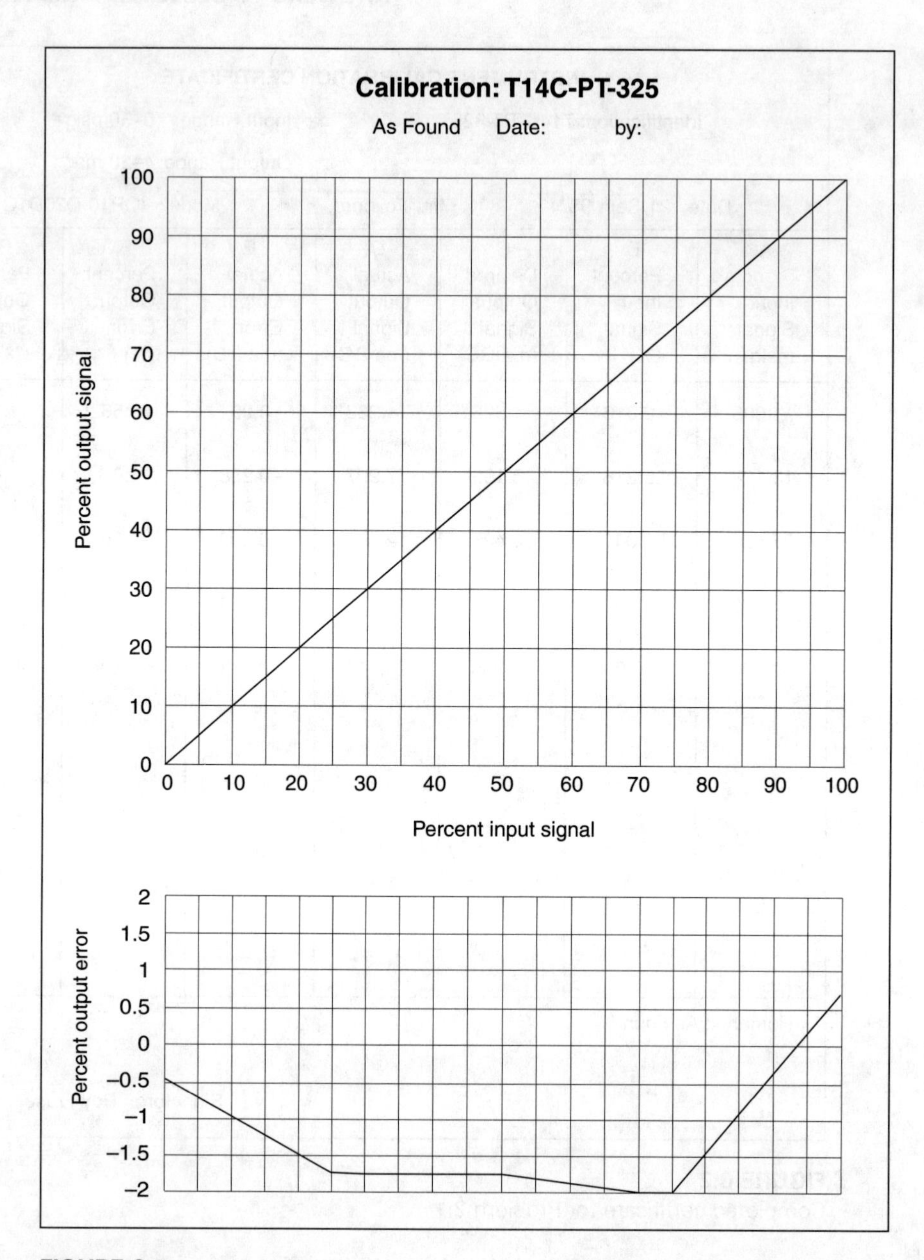

FIGURE C.3
Completed graphs for Problem 2.1

Problem 2.10

Fiber-optic infrared industrial temperature transmitter with a range of 800–1800°C and an output of 4–20 ma DC. This transmitter would be mounted on the stationary firing end of the kiln but protected from viewing the flame. It would be aimed at the molten cement that rolls down the inside of the kiln. Omega OS1553-A.

Problem 2.11

1 kPa = 0.145 psi = 4.015 in. of water.

–30 to +70 kPa gauge equals (–30 × 0.145) + 14.7 to (+70 × 0.145) + 14.7 or 10.35 to 24.85 psia.

–30 to +70 kPa gauge equals (–30 × 4.015) to (+70 × 4.015) or –120.45 to 281.05 in. of water gauge.

–30 to +70 kPa gauge equals (–30 × 0.145) to (+70 × 0.145) or – 4.35 to 10.15 psig.

Problem 2.13

Foxboro IGP10T20D1M2L1A1V3Z3C2—Electronic Gauge Pressure Transmitter for Hart digital protocol plus 4–20 ma analog signal, with 316L stainless-steel process connection and silicone-fluid fill, span limits between 10–30 psi calibrated for a range of 0–40 ft of water, a ½″ NPT conduit connection, 304 ss mounting bracket, digital indicator, block and bleed valve 316 ss, external zero adjustment and custody transfer lock and seal, and full factory configuration.

Problem 2.14

Time	*lpm*	*Flow change*	*Liters since previous time*	*Total*
14:10:06	0	None	0	0
14:12:08	0	None	0	0
14:12:21	173	Ramp up	18.742 = 173 × 13/(60 × 2)	18.742
14:16:48	173	No change	769.85 = 173 × 4.45	788.59
14:16:55	41	Ramp down	12.483 = (173 + 41) × 7/(2 × 60)	801.08
14:17:23	0	Ramp down	9.567 = (41 + 0) × 28 / (2 × 60)	810.65

Problem 2.16

A vortex flowmeter places an obstruction in the stream. This obstruction deflects the flow around itself and generates vortices in the fluid. The slight variation in the fluid pressure due to each vortex is detected by a sensor in the vortex flowmeter. An electronic counter counts the rate of vortices per unit time and this is directly related to the fluid velocity. Multiplying the velocity by the cross-sectional area of the flowmeter produces the volumetric flow rate. The main advantage of the vortex flowmeter over the orifice flowmeter is that it produces a linear signal rather than a square root signal.

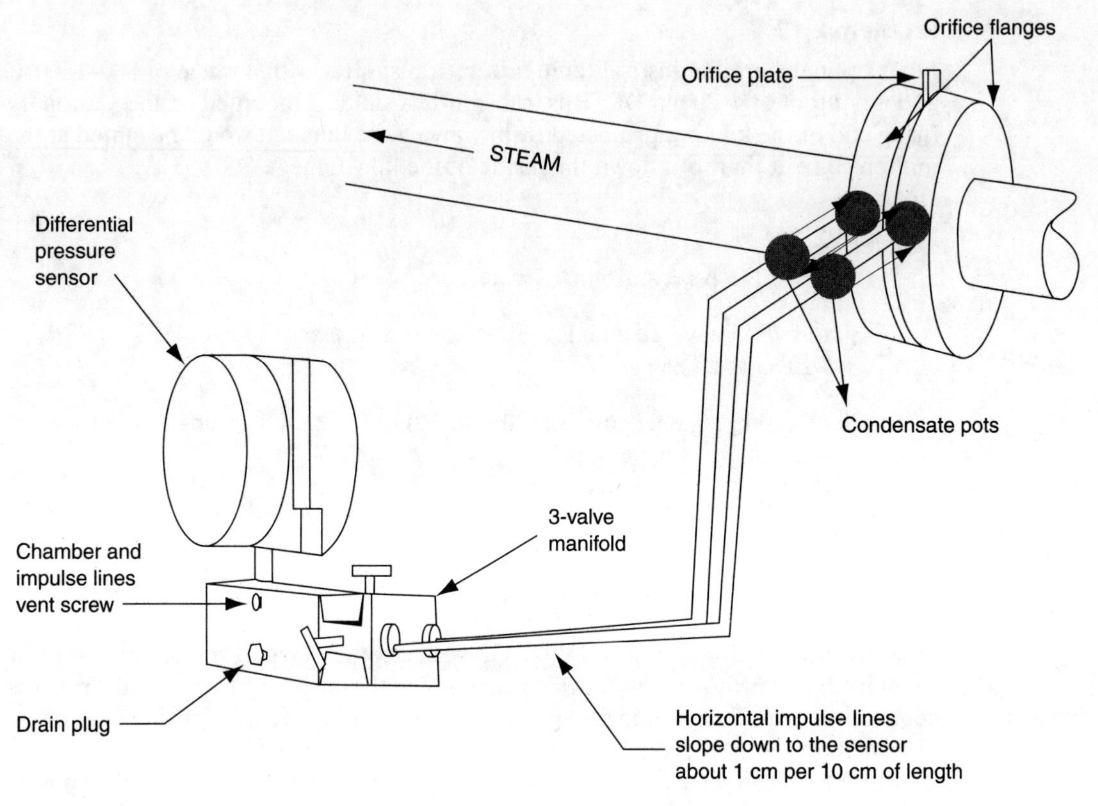

FIGURE C.4
Solution to Problem 2.19

Problem 2.19

The solution to this problem is shown in Figure C.4.

Problem 2.20

Reject ultrasonic doppler—no bubbles.

Reject magnetic—conductivity too low.

Reject turbine—mechanical maintenance required.

Orifice plate may be correct, but square root signal is undesirable.

Ultrasonic transit time correct, but I prefer vortex flowmeter.

Problem 2.21

14.38 ma = (14.38 – 4) × 100 / 16 = 64.875 kPa = 64.875 / (9.80665 × 1.0375) = 6.3763 m = 637.63 cm + 10 cm = 647.63 cm above the tank bottom. Span is 100 kPa or 982.859 cm. Maximum expected error is 0.25 × 982.859 / 100% = 2.46 cm.

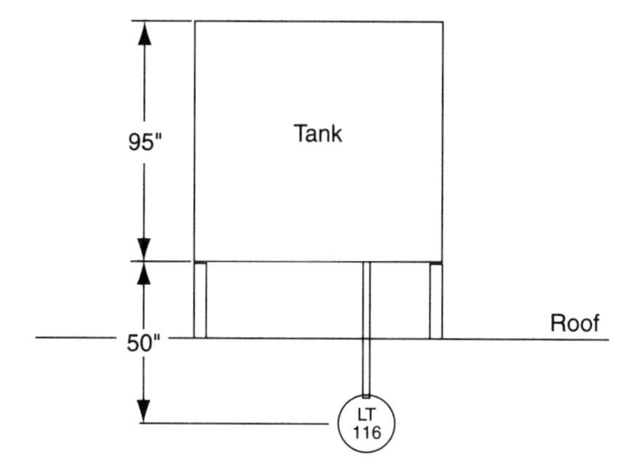

FIGURE C.5
Solution to Problem 2.24

Problem 2.24

See Figure C.5. Zero signal (4 ma) occurs at 50 in. of water. The span should be 95 in. of water. The range becomes 50–145 in. of water. The zero is suppressed since the process variable value of 0 in. of water pressure is below the 4 ma value.

Problem 2.25

(a) An ultrasonic level meter requires little maintenance to measure flour level in a silo. Flour dust in the air is a possible problem, but 30 meters will probably be satisfactory. Silo weight is another method, but it is expensive to use load cells for such a heavy structure, and maintenance on them would be expensive.

(b) A paddle-wheel switch with a shaft 1 m long is an effective device for detecting cement near the roof of a silo. If it stops for any reason, it trips off the conveyor. It is cheaper and simpler than an ultrasonic device or a radioactive device.

(c) A differential pressure sensor is effective for level measurement of a foamy, turbulent liquid. A capacitance sensor and an ultrasonic sensor are much affected by foam, and a displacer is affected by turbulence.

Problem 2.27

From Figure 2.16 the dry bulb 72.3°F vertical line crosses the wet bulb 55.5°F at the relative humidity line of about 32%, the absolute humidity of 0.0054 lb of moisture per pound of dry air, dew point of 41°F.

Problem 2.29

From Figure 2.17 for a dew point of 40°F, the ambient temperature inside the DEWCEL is 107°F. For 41°F dew point, the DEWCEL is 108°F.

Problem 2.31

$$G = \gamma A / l$$

for $G = 300$ μsiemens, $l = 2$ cm, $A = 4$ cm^2,
and then $\gamma = 300 \times 2 / (2 \times 2) = 150$ μsiemens/cm.

Problem 2.33

2500 hp to the load requires $2500 \times 0.746 / 0.97 = 1923$ kW.

1923 kW requires $1923 / 0.9 = 2136$ kVA.

2136 kVA draws $2136 / (1.732 \times 4.300) = 287$ A in each line.

4–20 ma corresponds to [–1000 W to 0 W to +1000 W] × (4800 / 120) × (300 / 5) = –2,400,000 W to 0 W to +2,400,000 W. See Figure C.6.

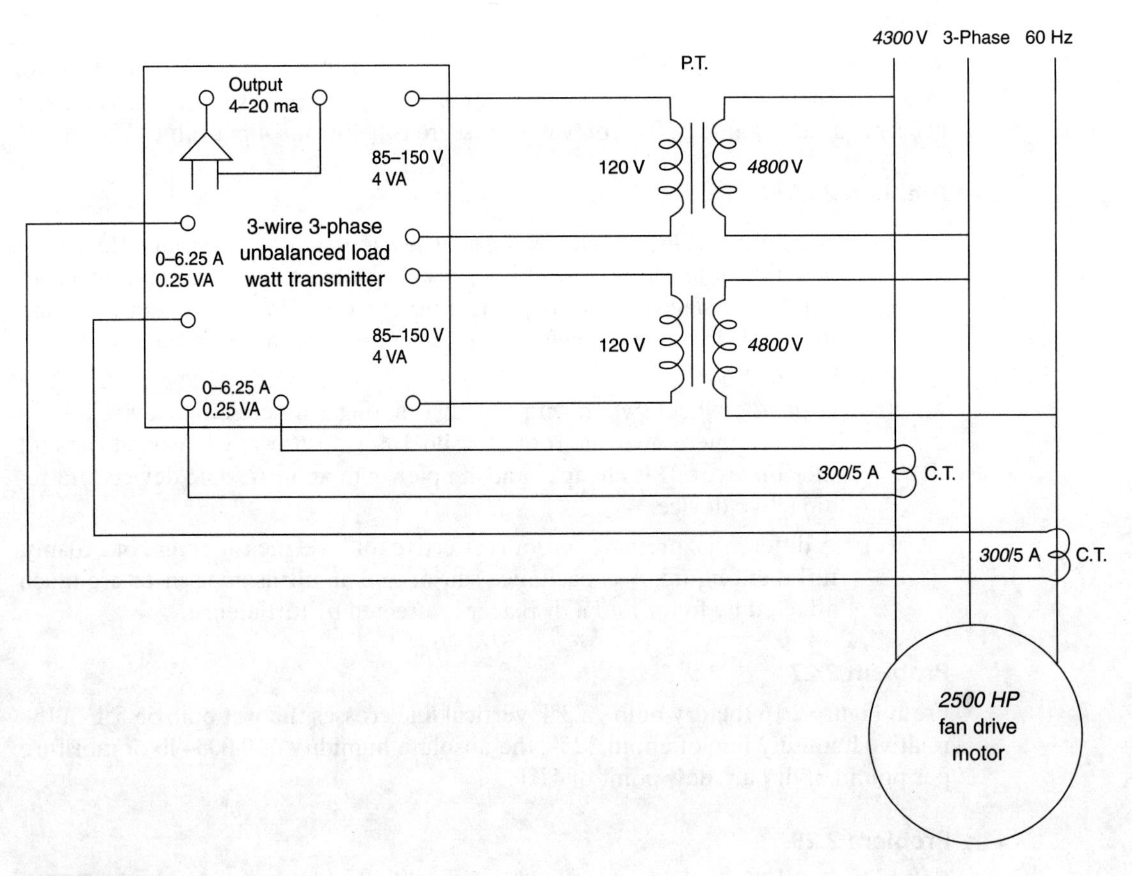

FIGURE C.6
Diagram for Problem 2.33

Problem 2.36

Frequency = 0.3927 rad/sec (2π radians or 360° for 16 secs).

Phase shift = 1 time div from peak of output channel to peak of input channel, which is 1 div × 360° / 8 div = –45°.

Channel 2 peak-to-peak = (4 voltage divs × 0.5 V/div) / 250 Ω = 8 ma_{pp}.

Channel 1 peak-to-peak = (5 voltage divs × 1 V/div) / 500 Ω = 10 ma_{pp}.

Gain = output / input = 8 / 10 = 0.8 ma/ma.

CHAPTER 3—PRESENTATION OF DATA

Problem 3.1

From 14:28 to 14:31 there are 8 dots. Ten dots will occur in 3 × 60 × 10 / 8 = 225 seconds. One cycle in 225 seconds is 0.0044 hertz.

Problem 3.2

For the second loop, with the name T12, *SP*/*Y* is shown as 68.7 for the set-point value. *PV*/*X* is shown as 72.1 for the process variable value. The upper loop descriptor of the second loop shows DEGC for the loop bearing the loop name T12. Since T12 is an ISA loop designation for temperature, then DEGC (from intuition) must be degrees Celsius. For the fourth loop, L5, the output signal has a value of 76.9 in units of percent. Its control status is *M* for manual. Its ISA tag is L5, a level loop.

Problem 3.3

The solution for this problem is shown in Figure C.7.

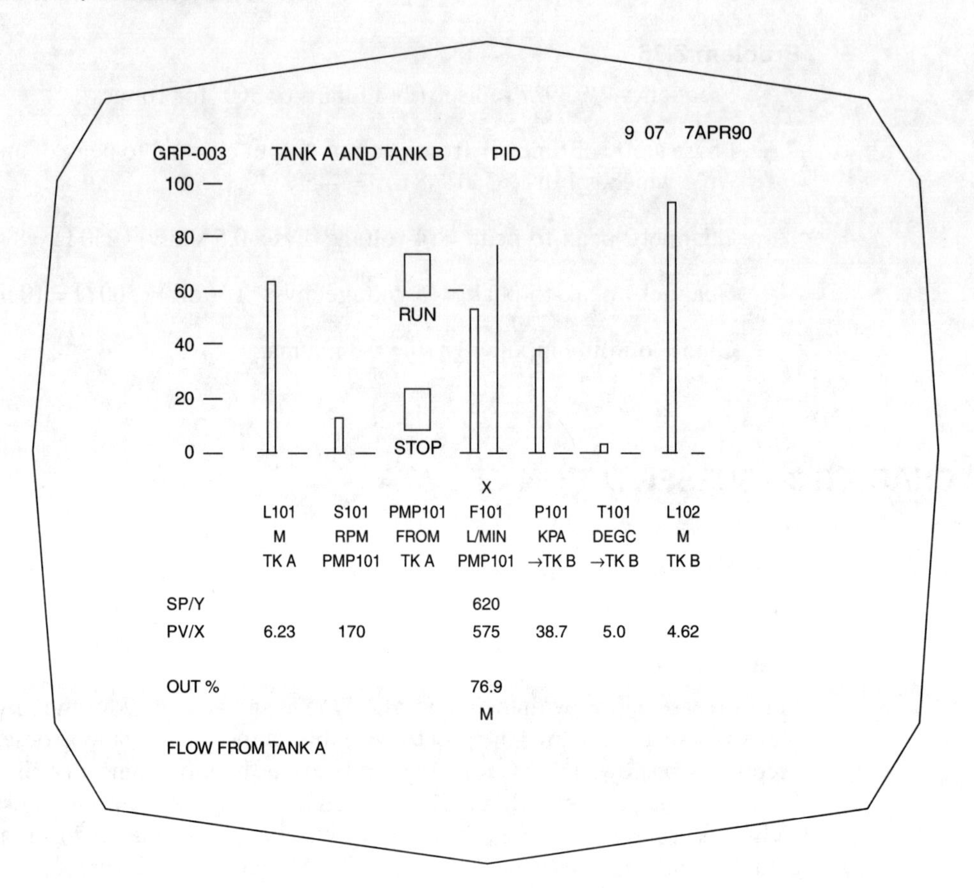

FIGURE C.7
Solution to Problem 3.3—Honeywell TDC 3000

CHAPTER 4—DATA COMMUNICATIONS

Problem 4.1

To find an undesired (and as yet undiscovered) ground on the shield at TT-105 with 65 Ω to ground at terminal 145 in terminal box I-1 with the wire to the ground electrode removed from terminal 145:

1. Temporarily disconnect the grounding strap from terminal 143 to terminal 145. Is the ground still on terminal 145? Yes. Put the grounding strap back on terminal 143.
2. Lift the shield connection to terminal 145. Is terminal 145 still grounded? No. Put the shield back on terminal 145. The ground is on the shield somewhere.

3. Go to terminal box I-17, and lift the shield connection for cable I-27 to terminal 79. Check by radio or phone if the ground is still on terminal 145 in box I-1. The answer is No. Put back the shield to 79.

4. Lift the wire from terminal 63 to terminal 66 in box I-17. Is the ground still on 145 in I-1? No. It must be on at least one of the shields of LT-21, FT-198, PT-217, or TT-105. Replace the lifted wire. Lift the wire from 35 to 32. Is the ground still on 145? No. It must be on PT-217 or TT-105. Replace the lifted wire. Lift it at 29, and again it is gone. It must be on the shield of TT-105. Replace the lifted wire. Examine the shield near terminals 27 and 28 and at the sensor TT-105 for an inadvertent ground. Remove it and reconnect all wires and check that terminal 145 in box I-1 is free of grounds. Reconnect the ground electrode to terminal 145. It should have been possible to continue acquiring good data all the time that these tests were being made. The operators, however, should have been alerted to the possibility of finger trouble from the service technician.

To find an undesired ground on the shield and negative terminal of PT-217, lift the shield from 145 and find that the ground does not disappear, but the resistance increases from 65 Ω to 400 Ω. Put the shield back on 145 and lift the ground strap to 143 and find that the ground does not disappear, but returns to 70 Ω. It is on both the shield and the signal circuits. Advise the operators that the ground exists on both circuits and obtain approval to temporarily shut down data acquisition from some of the sensors. Remove the shield from 145, put back the wire from 145 to 143, and lift the wire from 132 to 135. The ground resistance increases from 400 Ω to 750 Ω at terminal 145 because the ground now travels through the 24 V power supply and the loop circuit. Put back the wire from 132 to 135 and disconnect LT-21 from 130 and 131. The resistance at 145 remains at 750 Ω. Reconnect LT-21 and disconnect FT-198 and again 750 Ω. Repeat for PT-217 and this time the resistance goes to 3 mΩ. The problem is at PT-217. Reconnect all wires and go to box I-17. Disconnect shield and signal wires of PT-217 and check at 145 for ground resistance. It is still at 3 mΩ. Find the offending grounds, remove them, make sure all circuits are functioning correctly, and advise the operators that the system is functioning correctly.

Problem 4.2

Tubing (ft)	*Dead Time (sec)*	*Five Time Constants (sec)*	*Sum (sec)*
100	0.108	0.75	0.86
200	0.2	2.0	2.2
300	0.3	4.0	4.3
400	0.4	6.0	6.4
500	0.5	8.5	9.0

Problem 4.4

The solution for this problem is shown in Figure C.8.

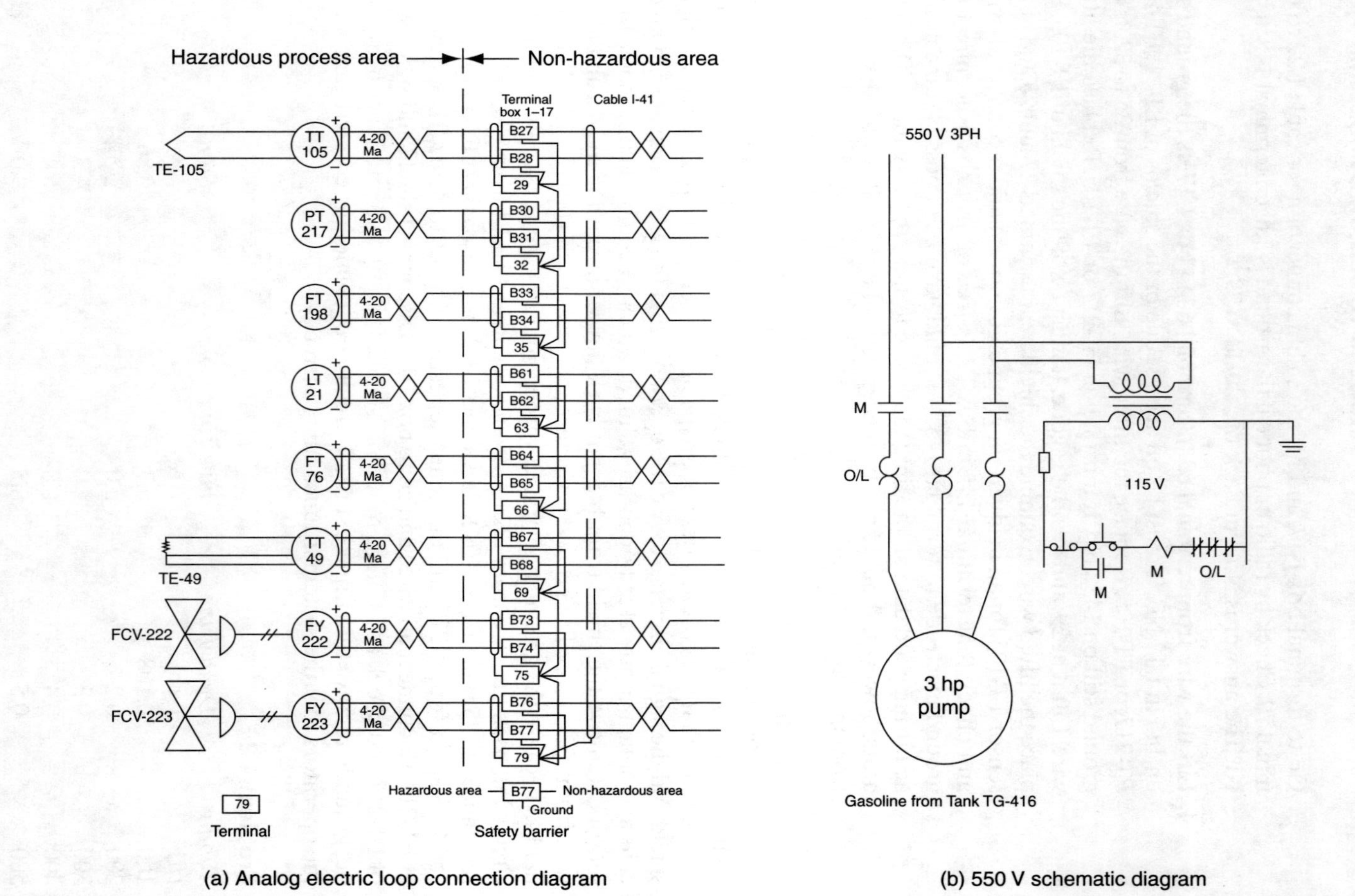

(a) Analog electric loop connection diagram

(b) 550 V schematic diagram

FIGURE C.8
Solution to Problem 4.4

Problem 4.6

4 + 4 bytes to request and receive each update per sensor. For 15 sensors this is 8 × 15 = 120 bytes. If each byte uses 10 bits, then this is 10 × 120 = 1200 bits per 15 sensors. At 9600 baud then, this provides 1200 bits per 15 sensors / 9600 bps = 1/8 seconds per 15 sensors = 0.125 seconds per 15 sensors to read all the sensors once.

CHAPTER 5—PROCESS ADJUSTMENT

Problem 5.2

	Water Temperature		
	60°F	*40°F*	*140°F*
1 cu ft	62.344	62.422	61.387
G	1.00000	1.0013	0.98465

C_v (40°F) = 160 × sqr(1.0013 / 14) = 42.790.

C_v (140°F) = 160 × sqr(0.98465 / 14) = 42.432.

Change in required C_v = –0.00042 or –0.00097% as water temperature increases from 40°F to 140°F.

Approximate valve size required is 2″ with C_v = 60.

Problem 5.3

First try: Assume 700 GPM with ΔP = 42 – 25 psi = 17 psid.

C_v = 700 × sqr(0.83 / 17) = 154.67 (too small)

Second try: Assume 800 GPM with ΔP = 36 – 31 psi = 5 psid.

C_v = 800 × sqr(0.83 / 5) = 325.94 (still slightly small)

Third try: Assume 820 GPM with ΔP = 35.5 – 31.7 psi = 3.8 psid

C_v = 820 × sqr(0.83 / 3.8) = 383 (close to 372)

Fourth try: Maximum flow is a little less than 820 GPM for C_v of 372, say about 815 GPM.

Minimum flow = (372 / 50) × sqr[(55 – 5) / 0.83] = 57.74 GPM

It is fairly certain that this valve will smoothly adjust the flow from 60–800 GPM.

Problem 5.7

20 ma has a C_v of 20. 4 ma has a C_v of 20 exp(0) / 35 = 0.5714.

A valve signal of 10.8 ma is (10.8 – 4) × 100% / (20 – 4) = 42.5% open.

C_{vp} = 20 × {exp[(42.5 / 100) × loge(35)]} / 35 = 2.589

Problem 5.8

Percent valve opening	5	10	20	40	60	100
C_{vp}	4	6	8	17	27	100
First try ΔP	50	50	49	46	35	35
First try flow	50	49	56	100	250	250
First try C_v	7	7	8	15	42	42
Second try ΔP	52	51		45	40	18
Second try flow	30	43		115	200	350
Second try C_v	4	6		17	32	82
Third try ΔP					42	5
Third try flow					175	400
Third try C_v					27	178
Fourth try ΔP						14
Fourth try flow						370
Fourth try C_v						99

Flow	30	43	56	115	175	370
Percent valve opening	5	10	20	40	60	100

Notice that the C_v line in Figure C.9 is much more curved than the flow line, as one might expect from an equal percentage valve.

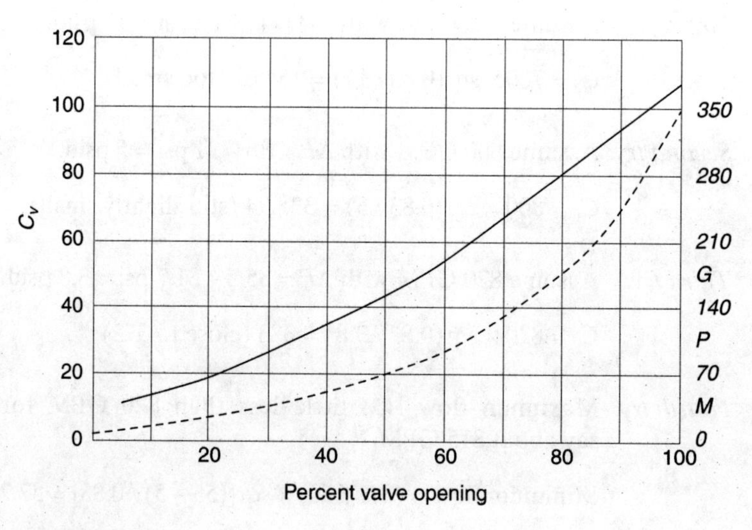

FIGURE C.9
Solution to Problem 5.8

Problem 5.10

From Figure 5.7 at a speed that is 70% rated speed, the pressure rise across the fan, curve *C*, will be 0.7 × 0.7 = 0.49 times the pressure rise at full speed. At very low flows, this will be approximately 500 × 0.49 = 245 Pa. The minimum flow will be 514 lps × sqr(0.245 / 0.505) = 358 lps. At 5500 lps and 70% speed, the pressure rise in the fan is 430 × 0.49 = 210 Pa and the pressure drop across the damper is 210 – 125 = 85 Pa, so the flow is 7523 lps × sqr(0.85 / 0.120) = 6331, try again. At 5800 lps the supply curve is 420 × 0.49 = 206 and the demand curve is 130 Pa. So the flow is 7523 lps × sqr[(0.206 – 0.130) / 0.120] = 5987. Estimate the maximum flow at 5900 lps.

Problem 5.12

Using equation 5.3 on page 119,

$$C_v = 20,000 \times \frac{\sqrt{\dfrac{28.97 \times 520 \times 1}{14.7 \times 0.04}}}{7320 \times 1}$$

Choose a 4″ valve with a $C_v = 500$.

Maximum flow = 20,000 × 500 / 437.33 = 22,866 SCFH

$$\text{Minimum flow} = 500 \times 7320 \times \frac{\sqrt{\dfrac{14.7 \times 0.2}{28.97 \times 520}}}{35} = 1461 \text{ SCFH}$$

This valve should readily adjust the flow from 2000 to 20,000 SCFH.

CHAPTER 6—AUTOMATIC CONTROL

Problem 6.2

Gain = 100% / 20% = 5. The solution to the block diagram is shown in Figure C.10.

Controller Output Signal	*Error*	F_B
10%	–8%	23%
50%	0%	15%
80%	6%	9%

This offset error can be eliminated by adding integral action to the controller.

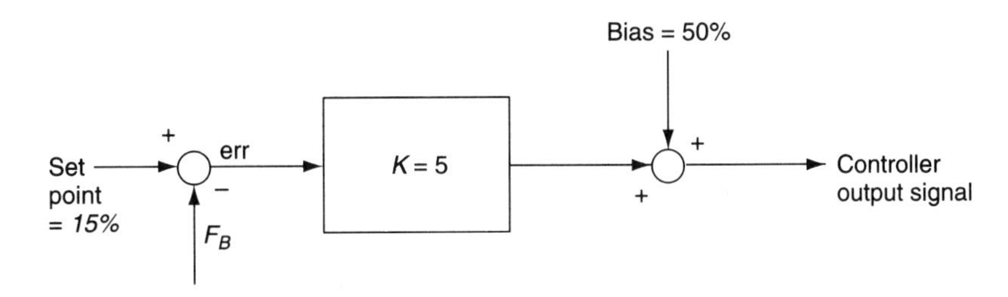

FIGURE C.10
Solution to Problem 6.2

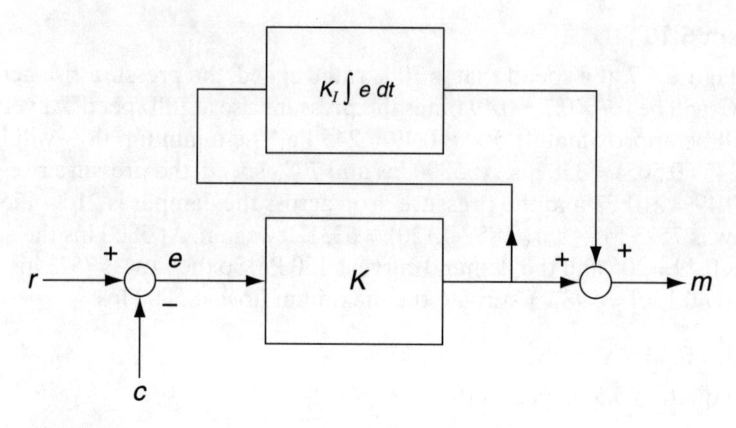

FIGURE C.11
Solution to Problem 6.5

Problem 6.4

$C_O = (S_P - F_B) \times K + \text{bias} = (33 - 35.5) \times 4 + 50 = 40\%$

Problem 6.5

To automatically obtain error, $e = 0\%$, you must add integral action to the proportional controller of Problem 6.1 as shown in Figure C.11.

The revised controller shown in Figure C.11 has integral action added to the proportional action. If the error, e is not 0, then the output from the proportional block with gain K will not be 0, and thus the K_I block will have an input that is not 0. If the input to K_I is positive, then it will continually ramp up its output. If negative, it will ramp down. The output signal will adjust the process and thus feed back a signal to point *c*. As the value of the *c* signal gets closer to the set point, *r*, then the value of error, *e*, will get closer to 0, and the rate of ramping by K_I will decrease. When it reaches zero, then ramping will cease and K_I will simply hold constant its last value. Therefore the offset error becomes 0.

Problem 6.6

The solution for this problem is shown in Figure C.12.

The overall open loop gain = $X \times 3$ (GPM / %) $\times$ 0.25 (% / GPM) = 0.5. Solving for *X*:

$$X = \frac{0.5}{3 \times 0.25} = 0.66667$$

To obtain an overall loop gain of 0.5 and consequently the quarter-amplitude response, set the controller gain to 0.66667 or a proportional band of 150%.

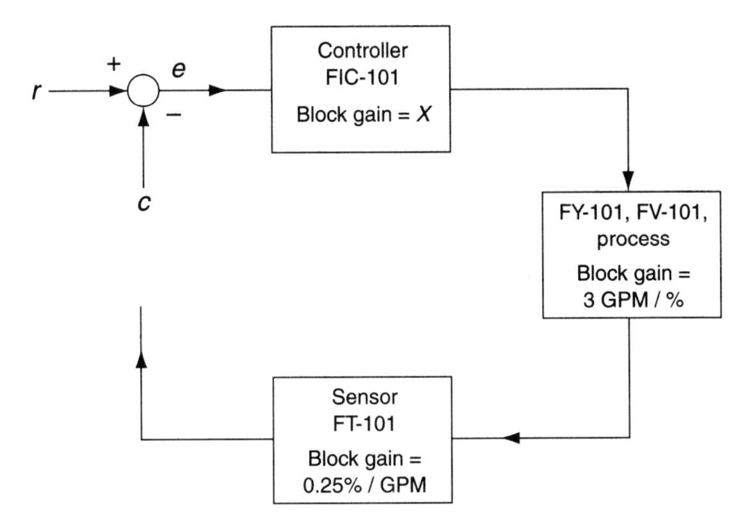

FIGURE C.12
Solution to Problem 6.6

Problem 6.7

With an overall open loop gain of 2 and a phase shift of –180° at the resonant or natural frequency of oscillation, then the closed loop response will be the undesirable response of an expanding amplitude of oscillation. In order to achieve quarter-amplitude response, the overall open loop gain should be 0.5. The controller proportional gain needs to be cut to one-quarter of the gain that it had for an open loop gain of 2. To achieve critical damping, the controller gain needs to be decreased from the value for quarter-amplitude response.

Problem 6.8

The solution to this problem is shown in Figure C.13.

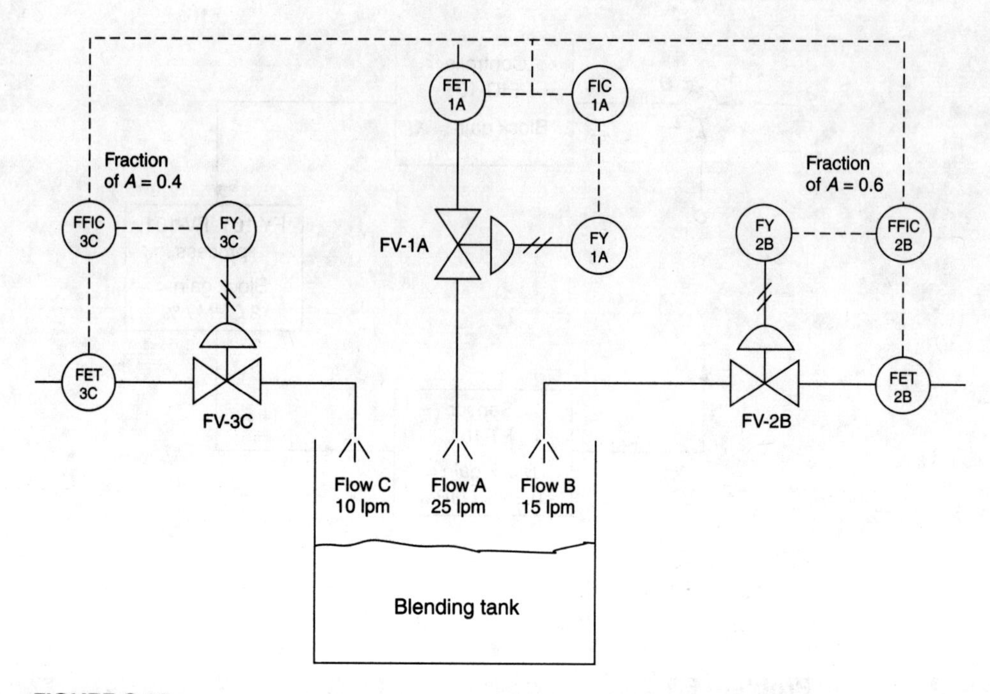

FIGURE C.13
Solution to Problem 6.8

Appendix D

ASCII SYMBOLS AND CODES

Decimal	Hex	^C key	Telecom Control Function
0	00	@	NUL (used for padding)
1	01	A	SOH (start of header)
2	02	B	STX (start of text)
3	03	C	ETX (end of text)
4	04	D	EOT (end of transmission)
5	05	E	ENQ (enquiry)
6	06	F	ACK (acknowledge)
7	07	G	BEL (bell)
8	08	H	BS (backspace)
9	09	I	HT (horizontal tab)
10	0A	J	LF (line feed)
11	0B	K	VT (vertical tab)
12	0C	L	FF (form feed or new page)
13	0D	M	CR (carriage return)
14	0E	N	SO (shift out)
15	0F	O	SI (shift in)
16	10	P	DLE (data link escape)
17	11	Q	DC1 (device control 1, XON)
18	12	R	DC2 (device control 2)
19	13	S	DC3 (device control 3, XOFF)
20	14	T	DC4 (device control 4)
21	15	U	NAK (negative acknowledge)
22	16	V	SYN (synchronous idle)
23	17	W	ETB (end transmission block)
24	18	X	CAN (cancel)
25	19	Y	EM (end of medium)
26	1A	Z	SUB (substitute)
27	1B	[	ESC (escape)
28	1C	\	FS (file separator)
29	1D	]	GS (group separator)
30	1E	^	RS (record separator)
31	1F	_	US (unit separator)

Decimal	Hex	Screen Char	Decimal	Hex	Screen Char
32	20	(space)	64	40	@
33	21	!	65	41	A
34	22	"	66	42	B
35	23	#	67	43	C
36	24	$	68	44	D
37	25	%	69	45	E
38	26	&	70	46	F
39	27	'	71	47	G
40	28	(	72	48	H
41	29	)	73	49	I
42	2A	*	74	4A	J
43	2B	+	75	4B	K
44	2C	,	76	4C	L
45	2D	–	77	4D	M
46	2E	.	78	4E	N
47	2F	/	79	4F	O
48	30	0	80	50	P
49	31	1	81	51	Q
50	32	2	82	52	R
51	33	3	83	53	S
52	34	4	84	54	T
53	35	5	85	55	U
54	36	6	86	56	V
55	37	7	87	57	W
56	38	8	88	58	X
57	39	9	89	59	Y
58	3A	:	90	5A	Z
59	3B	;	91	5B	[
60	3C	<	92	5C	\
61	3D	=	93	5D	]
62	3E	>	94	5E	^
63	3F	?	95	5F	_

<table>
<tr><th>Decimal</th><th>Hex</th><th>Screen Char</th><th>Decimal</th><th>Hex</th><th>Screen Char</th></tr>
<tr><td>96</td><td>60</td><td>`</td><td>128</td><td>80</td><td>Ç</td></tr>
<tr><td>97</td><td>61</td><td>a</td><td>129</td><td>81</td><td>ü</td></tr>
<tr><td>98</td><td>62</td><td>b</td><td>130</td><td>82</td><td>é</td></tr>
<tr><td>99</td><td>63</td><td>c</td><td>131</td><td>83</td><td>â</td></tr>
<tr><td>100</td><td>64</td><td>d</td><td>132</td><td>84</td><td>ä</td></tr>
<tr><td>101</td><td>65</td><td>e</td><td>133</td><td>85</td><td>à</td></tr>
<tr><td>102</td><td>66</td><td>f</td><td>134</td><td>86</td><td>å</td></tr>
<tr><td>103</td><td>67</td><td>g</td><td>135</td><td>87</td><td>ç</td></tr>
<tr><td>104</td><td>68</td><td>h</td><td>136</td><td>88</td><td>ê</td></tr>
<tr><td>105</td><td>69</td><td>i</td><td>137</td><td>89</td><td>ë</td></tr>
<tr><td>106</td><td>6A</td><td>j</td><td>138</td><td>8A</td><td>è</td></tr>
<tr><td>107</td><td>6B</td><td>k</td><td>139</td><td>8B</td><td>ï</td></tr>
<tr><td>108</td><td>6C</td><td>l</td><td>140</td><td>8C</td><td>î</td></tr>
<tr><td>109</td><td>6D</td><td>m</td><td>141</td><td>8D</td><td>ì</td></tr>
<tr><td>110</td><td>6E</td><td>n</td><td>142</td><td>8E</td><td>Ä</td></tr>
<tr><td>111</td><td>6F</td><td>o</td><td>143</td><td>8F</td><td>Å</td></tr>
<tr><td>112</td><td>70</td><td>p</td><td>144</td><td>90</td><td>É</td></tr>
<tr><td>113</td><td>71</td><td>q</td><td>145</td><td>91</td><td>æ</td></tr>
<tr><td>114</td><td>72</td><td>r</td><td>146</td><td>92</td><td>Æ</td></tr>
<tr><td>115</td><td>73</td><td>s</td><td>147</td><td>93</td><td>ô</td></tr>
<tr><td>116</td><td>74</td><td>t</td><td>148</td><td>94</td><td>ö</td></tr>
<tr><td>117</td><td>75</td><td>u</td><td>149</td><td>95</td><td>ò</td></tr>
<tr><td>118</td><td>76</td><td>v</td><td>150</td><td>96</td><td>û</td></tr>
<tr><td>119</td><td>77</td><td>w</td><td>151</td><td>97</td><td>ù</td></tr>
<tr><td>120</td><td>78</td><td>x</td><td>152</td><td>98</td><td>ÿ</td></tr>
<tr><td>121</td><td>79</td><td>y</td><td>153</td><td>99</td><td>Ö</td></tr>
<tr><td>122</td><td>7A</td><td>z</td><td>154</td><td>9A</td><td>Ü</td></tr>
<tr><td>123</td><td>7B</td><td>{</td><td>155</td><td>9B</td><td>¢</td></tr>
<tr><td>124</td><td>7C</td><td>|</td><td>156</td><td>9C</td><td>£</td></tr>
<tr><td>125</td><td>7D</td><td>}</td><td>157</td><td>9D</td><td>¥</td></tr>
<tr><td>126</td><td>7E</td><td>~</td><td>158</td><td>9E</td><td>₧</td></tr>
<tr><td>127</td><td>7F</td><td>ã</td><td>159</td><td>9F</td><td>ƒ</td></tr>
</table>

127_{10} ã = DEL (delete) or
ASCII 008 (back space) in DOS

Decimal	Hex	Screen Char	Decimal	Hex	Screen Char
160	A0	á	192	C0	└
161	A1	í	193	C1	┴
162	A2	ó	194	C2	┬
163	A3	ú	195	C3	├
164	A4	ñ	196	C4	─
165	A5	Ñ	197	C5	┼
166	A6	ª	198	C6	╞
167	A7	º	199	C7	╟
168	A8	¿	200	C8	╚
169	A9	⌐	201	C9	╔
170	AA	¬	202	CA	╩
171	AB	½	203	CB	╦
172	AC	¼	204	CC	╠
173	AD	¡	205	CD	═
174	AE	«	206	CE	╬
175	AF	»	207	CF	╧
176	B0	░	208	D0	╨
177	B1	▒	209	D1	╤
178	B2	▓	210	D2	╥
179	B3	│	211	D3	╙
180	B4	┤	212	D4	╘
181	B5	╡	213	D5	╒
182	B6	╢	214	D6	╓
183	B7	╖	215	D7	╫
184	B8	╕	216	D8	╪
185	B9	╣	217	D9	┘
186	BA	║	218	DA	┌
187	BB	╗	219	DB	█
188	BC	╝	220	DC	▄
189	BD	╜	221	DD	▌
190	BE	╛	222	DE	▐
191	BF	┐	223	DF	▀

Decimal	Hex	Screen Char
224	E0	α
225	E1	ß
226	E2	Γ
227	E3	π
228	E4	Σ
229	E5	σ
230	E6	µ
231	E7	τ
232	E8	Φ
233	E9	Θ
234	EA	Ω
235	EB	δ
236	EC	∞
237	ED	φ
238	EE	ε
239	EF	∩
240	F0	≡
241	F1	±
242	F2	≥
243	F3	≤
244	F4	⌠
245	F5	⌡
246	F6	–
247	F7	≈
248	F8	°
249	F9	•
250	FA	
251	FB	√
252	FC	^{n}
253	FD	^{2}
254	FE	■
255	FF	

Appendix E

PIPING AND FITTINGS

The pressure drop in a pipe is like the voltage drop in an electric resistor. But there are two regimes of this pressure drop. For Reynolds numbers below 2000, the pressure drop is linearly proportional to the flow rate. This is called *laminar flow.* For Reynolds numbers above 10,000, the pressure drop is proportional to the square of the flow rate. This is called *turbulent flow.* From 2000 to 10,000 is a transition region. The voltage drop is simpler in that it is linearly proportional to the electric current flow rate over the whole range of current flow rate.

Reynolds number is defined as

$$R_e = \frac{\rho \nu d}{\mu}$$

where: ρ = density of the fluid

ν = average velocity of flowing fluid

d = inside diameter of pipe

μ = absolute viscosity of the fluid

For practical use of pipes to transfer fluids from one location to another, the flow rates are almost always turbulent, with Reynolds numbers varying from 10,000 to 10,000,000. Notice that the velocity multiplied by the diameter squared is proportional to the volumetric flow rate, and this multiplied by the density is proportional to the mass flow rate, W. Therefore, the pipe Reynolds number may also be calculated as follows:

$$R_D = \frac{W}{235.6 \times D \times \mu \times g_c}$$

where: W = mass flow rate, lb_m/hour

D = pipe inside diameter, inches

μ = absolute viscosity, $lb_f\text{-sec}/ft^2$

g_c = units constant, 32.2 $lb_m\text{-ft}/(lb_f\text{-sec}^2)$

For SI units this becomes

$$R_D = \frac{W}{0.787 \times D \times \mu \times g_c}$$

where: W = mass flow rate, kg/sec
D = pipe inside diameter, m
μ = absolute viscosity, Pa-seconds
g_c = units constant, 1.0 kg_m-m/(N-sec^2)

Notice that all the units in these formulas cancel out, and for this reason the formulas are dimensionless.

The Fanning equation relates pressure drop over the length, l, of the pipe. This equation is as follows:

$$p = \frac{8\,\rho\, f\, l\, Q^2}{\pi^2\, d^5}$$

where: Q = flow rate, m^3/sec
ρ = fluid density, kg/m^3
f = friction factor
l = length of pipe, m
d = inside diameter of pipe, m
p = pressure drop over length l, Pa

The friction factor depends slightly on the Reynolds number, on the smoothness of the inside surface of the pipe, and on the diameter of the pipe. Computer programs exist to calculate the pressure drop, such as LIQRESIS.*

Values of the friction factor are shown in Table E.1.

EXAMPLE E.1 **Pressure Drop in 50 m of 10 cm SCH 40 Pipe**

Fifty meters of 10 cm (4 in.) schedule 40 commercial steel pipe, with an actual inside diameter of 10.226 cm (4.026 in.) has water at 5°C flowing through it at a flow rate of 0.02464 m^3/sec (velocity of 3 m/sec or 9.84 ft/sec). What is the pressure drop in this pipe? The Reynolds number is 24.64 (kg/sec) / [0.787 × 0.10226 (m) × 0.001 (Pa-sec) × 1.0 (kg_m-m/(N-sec^2))] = 306,168. The friction factor is 0.020. The pressure drop is 8 × 1000 (kg/m^3) × 0.020 × 50 (m) × [0.02464 (m^3/sec)]^2 / (9.8696 × [0.10226 (m)]^5 = 44010 Pa, or 6.378 psi, or 3.89 psi for 100 ft of pipe.

Pipe fittings may also cause significant pressure loss. Their contribution to the pressure loss is accounted for by adding on an equivalent length of pipe to the overall length of bare pipe. Manufacturers of pipe fittings can provide their equivalent lengths.

*See Appendix F of R. Bateson. 1996. *Introduction to Control System Technology,* 5th ed. Upper Saddle River, NJ: Prentice Hall.

TABLE E.1
Values of friction factor

Type	Diameter (cm)	Reynolds Number				
		10^4	10^5	10^6	10^7	10^8
Smooth tubing	1–2	0.030	0.018	0.014	0.012	0.012
	2–4	0.030	0.018	0.013	0.011	0.010
	4–8	0.030	0.018	0.012	0.010	0.009
	8–16	0.030	0.018	0.012	0.009	0.008
Commercial steel pipe	1–2	0.035	0.028	0.026	0.026	0.026
	2–4	0.033	0.024	0.023	0.023	0.023
	4–8	0.030	0.022	0.020	0.019	0.019
	8–16	0.030	0.020	0.018	0.017	0.017

Source: L. F. Moody. 1944. Friction Factors for Pipe Flow. *ASME Transactions* 66 (4): 671.

Appendix F

THERMOCOUPLE AND RTD TABLES

TABLE F.1
Iron–constantan thermocouple type J, reference junction at 0°C

°C	Millivolts	°C	Millivolts	°C	Millivolts	°C	Millivolts	°C	Millivolts
–200	–7.89	–90	–4.215	20	1.019	130	6.907	240	12.996
–195	–7.778	–85	–4.001	25	1.277	135	7.182	245	13.275
–190	–7.659	–80	–3.785	30	1.536	140	7.457	250	13.553
–185	–7.533	–75	–3.566	35	1.797	145	7.732	255	13.83
–180	–7.402	–70	–3.344	40	2.058	150	8.008	260	14.108
–175	–7.265	–65	–3.12	45	2.321	155	8.284	265	14.385
–170	–7.122	–60	–2.892	50	2.585	160	8.56	270	14.663
–165	–6.974	–55	–2.663	55	2.849	165	8.837	275	14.94
–160	–6.821	–50	–2.431	60	3.115	170	9.113	280	15.217
–155	–6.663	–45	–2.197	65	3.381	175	9.39	285	15.494
–150	–6.499	–40	–1.96	70	3.649	180	9.667	290	15.771
–145	–6.331	–35	–1.722	75	3.917	185	9.944	295	16.048
–140	–6.159	–30	–1.481	80	4.186	190	10.222	300	16.325
–135	–5.982	–25	–1.239	85	4.455	195	10.499	305	16.602
–130	–5.801	–20	–0.995	90	4.725	200	10.777	310	16.879
–125	–5.615	–15	–0.748	95	4.996	205	11.054	315	17.155
–120	–5.426	–10	–0.501	100	5.268	210	11.332	320	17.432
–115	–5.233	–5	–0.251	105	5.54	215	11.609	325	17.708
–110	–5.036	0	0	110	5.812	220	11.887	330	17.984
–105	–4.836	5	0.253	115	6.085	225	12.165	335	18.26
–100	–4.632	10	0.507	120	6.359	230	12.442	340	18.537
–95	–4.425	15	0.762	125	6.633	235	12.72	345	18.813

(Continued on next page)

TABLE F.1
(Continued)

°C	Millivolts	°C	Millivolts	°C	Millivolts	°C	Millivolts	°C	Millivolts
350	19.089	480	26.272	610	33.683	740	41.647	870	49.989
355	19.364	485	26.551	615	33.977	745	41.965	875	50.305
360	19.64	490	26.829	620	34.273	750	42.283	880	50.621
365	19.916	495	27.109	625	34.569	755	42.602	885	50.936
370	20.192	500	27.388	630	34.867	760	42.922	890	51.249
375	20.467	505	27.668	635	35.165	765	43.242	895	51.562
380	20.743	510	27.949	640	35.464	770	43.563	900	51.875
385	21.019	515	28.23	645	35.764	775	43.885	905	52.186
390	21.295	520	28.911	650	36.066	780	44.207	910	52.496
395	21.57	525	28.793	655	36.368	785	44.529	915	52.806
400	21.846	530	29.075	660	36.671	790	44.852	920	53.115
405	22.122	535	29.358	665	36.975	795	45.175	925	53.422
410	22.397	540	29.642	670	37.28	800	45.498	930	53.729
415	22.673	545	29.926	675	37.586	805	45.821	935	54.035
420	22.949	550	30.21	680	37.893	810	46.144	940	54.341
425	23.225	555	30.496	685	38.201	815	46.467	945	54.645
430	23.501	560	30.782	690	38.51	820	46.79	950	54.948
435	23.777	565	31.068	695	38.819	825	47.112	955	55.251
440	24.054	570	31.356	700	39.13	830	47.434	960	55.553
445	24.33	575	31.644	705	39.442	835	47.755	965	55.854
450	24.607	580	31.933	710	39.754	840	48.076	970	56.155
455	24.884	585	32.222	715	40.068	845	48.397	975	56.454
460	25.161	590	32.513	720	40.382	850	48.716	980	56.753
465	25.438	595	32.804	725	40.697	855	49.036	985	57.051
470	25.716	600	33.096	730	41.013	860	49.354	990	57.349
475	25.994	605	33.389	735	41.329	865	49.672	995	57.646

TABLE F.2
Chromel–alumel thermocouple type K, reference junction at 0°C

°C	Millivolts	°C	Millivolts	°C	Millivolts	°C	Millivolts	°C	Millivolts
–200	–5.891	–25	–0.968	150	6.137	325	13.247	500	20.64
–195	–5.813	–20	–0.777	155	6.338	330	13.456	505	20.853
–190	–5.73	–15	–0.585	160	6.539	335	13.665	510	21.066
–185	–5.642	–10	–0.392	165	6.739	340	13.874	515	21.28
–180	–5.55	–5	–0.197	170	6.939	345	14.083	520	21.493
–175	–5.454	0	0	175	7.139	350	14.292	525	21.706
–170	–5.354	5	0.198	180	7.338	355	14.502	530	21.919
–165	–5.249	10	0.397	185	7.538	360	14.712	535	22.132
–160	–5.141	15	0.597	190	7.737	365	14.922	540	22.346
–155	–5.029	20	0.798	195	7.937	370	15.132	545	22.559
–150	–4.912	25	1	200	8.137	375	15.342	550	22.772
–145	–4.792	30	1.203	205	8.336	380	15.552	555	22.985
–140	–4.669	35	1.407	210	8.537	385	15.763	560	23.198
–135	–4.541	40	1.611	215	8.737	390	15.974	565	23.411
–130	–4.41	45	1.817	220	8.938	395	16.184	570	23.624
–125	–4.276	50	2.022	225	9.139	400	16.395	575	23.837
–120	–4.138	55	2.229	230	9.341	405	16.607	580	24.05
–115	–3.997	60	2.436	235	9.543	410	16.818	585	24.263
–110	–3.852	65	2.643	240	9.745	415	17.029	590	24.476
–105	–3.704	70	2.85	245	9.948	420	17.241	595	24.689
–100	–3.553	75	3.058	250	10.151	425	17.453	600	24.902
–95	–3.399	80	3.266	255	10.355	430	17.664	605	25.114
–90	–3.242	85	3.473	260	10.56	435	17.876	610	25.327
–85	–3.082	90	3.681	265	10.764	440	18.088	615	25.539
–80	–2.92	95	3.888	270	10.969	445	18.301	620	25.751
–75	–2.754	100	4.095	275	11.175	450	18.513	625	25.964
–70	–2.586	105	4.302	280	11.381	455	18.725	630	26.176
–65	–2.416	110	4.508	285	11.587	460	18.938	635	26.387
–60	–2.243	115	4.714	290	11.793	465	19.15	640	26.599
–55	–2.067	120	4.919	295	12	470	19.363	645	26.811
–50	–1.889	125	5.124	300	12.207	475	19.576	650	27.022
–45	–1.709	130	5.327	305	12.415	480	19.788	655	27.234
–40	–1.527	135	5.531	310	12.623	485	20.001	660	27.445
–35	–1.342	140	5.733	315	12.831	490	20.214	665	27.656
–30	–1.156	145	5.936	320	13.039	495	20.427	670	27.867

(Continued on next page)

TABLE F.2
(Continued)

°C	Millivolts	°C	Millivolts	°C	Millivolts	°C	Millivolts	°C	Millivolts
675	28.078	740	30.799	805	33.482	870	36.121	935	38.717
680	28.288	745	31.007	810	33.686	875	36.323	940	38.915
685	28.498	750	31.214	815	33.891	880	36.524	945	39.112
690	28.709	755	31.422	820	34.095	885	36.724	950	39.31
695	28.919	760	31.629	825	34.299	890	36.925	955	39.507
700	29.128	765	31.836	830	34.502	895	37.125	960	39.703
705	29.338	770	32.042	835	34.705	900	37.325	965	39.9
710	29.547	775	32.249	840	34.909	905	37.524	970	40.096
715	29.756	780	32.455	845	35.111	910	37.724	975	40.292
720	29.965	785	32.661	850	35.314	915	37.923	980	40.488
725	30.174	790	32.866	855	35.516	920	28.122	985	40.684
730	30.383	795	33.072	860	35.718	925	38.32	990	40.879
735	30.591	800	33.277	865	35.92	930	38.519	995	41.074

TABLE F.3
Platinum RTD 100 Ω at 0°C, DIN curve 43760, 9–68

°C	Ohms	°C	Ohms	°C	Ohms	°C	Ohms	°C	Ohms
–200	18.53	–40	84.21	120	146.06	280	204.88	440	260.75
–195	20.65	–35	86.19	125	147.94	285	206.68	445	262.45
–190	22.78	–30	88.17	130	149.82	290	208.46	450	264.14
–185	24.92	–25	90.15	135	151.7	295	210.25	455	265.83
–180	27.05	–20	92.13	140	153.57	300	212.03	460	267.52
–175	29.17	–15	94.1	145	155.45	305	213.81	465	269.21
–170	31.28	–10	96.07	150	157.32	310	215.58	470	270.89
–165	33.38	–5	98.04	155	159.18	315	217.36	475	272.57
–160	35.48	0	100	160	161.04	320	219.13	480	274.25
–155	37.57	5	101.95	165	162.9	325	220.9	485	275.92
–150	39.65	10	103.9	170	164.76	330	222.66	490	277.6
–145	41.73	15	105.85	175	166.62	335	224.42	495	279.27
–140	43.8	20	107.79	180	168.47	340	226.18	500	280.93
–135	45.87	25	109.73	185	170.32	345	227.94	505	282.6
–130	47.93	30	111.67	190	172.16	350	229.69	510	284.26
–125	49.99	35	113.61	195	174	355	231.44	515	285.91
–120	52.04	40	115.54	200	175.84	360	233.19	520	287.57
–115	54.09	45	117.47	205	177.68	365	234.93	525	289.22
–110	56.13	50	119.4	210	179.51	370	236.67	530	290.87
–105	58.17	55	121.32	215	181.34	375	238.41	535	292.51
–100	60.2	60	123.24	220	183.17	380	240.15	540	294.16
–95	62.23	65	125.16	225	185	385	241.88	545	295.8
–90	64.25	70	127.07	230	186.82	390	243.61	550	297.43
–85	66.27	75	128.98	235	188.64	395	245.34	555	299.07
–80	68.28	80	130.89	240	190.46	400	247.06	560	300.7
–75	70.29	85	132.8	245	192.27	405	248.78	565	302.33
–70	72.29	90	134.7	250	194.08	410	250.5	570	303.95
–65	74.29	95	136.6	255	195.89	415	252.21	575	305.58
–60	76.28	100	138.5	260	197.7	420	253.93	580	307.2
–55	78.27	105	140.39	265	199.5	425	255.64	585	308.81
–50	80.25	110	142.28	270	201.3	430	257.34	590	310.43
–45	82.23	115	144.18	275	203.09	435	259.05	595	312.04

(Continued on next page)

TABLE F.3
(Continued)

°C	Ohms	°C	Ohms	°C	Ohms	°C	Ohms	°C	Ohms
600	313.65	655	331.15	710	348.3	765	365.1	820	381.55
605	315.25	660	332.72	715	349.84	770	366.61	825	383.03
610	316.86	665	334.29	720	351.38	775	368.12	830	384.5
615	318.46	670	335.86	725	352.92	780	369.62	835	385.98
620	320.05	675	337.43	730	354.45	785	371.12	840	387.45
625	321.65	680	338.99	735	355.98	790	372.62	845	388.91
630	323.24	685	340.55	740	357.51	795	374.12	850	390.38
635	324.83	690	342.1	745	359.03	800	375.61		
640	326.41	695	343.66	750	360.55	805	377.1		
645	327.99	700	345.21	755	362.07	810	378.59		
650	329.57	705	346.76	760	363.59	815	380.07		

GLOSSARY

Arching In a silo, partially filled with solid particles, it is possible that, as particles are drawn out at the bottom, an arch of the particles may form across the top of the silo that will cause an incorrect reading of the level of the solid particles.

Baud Rate of transmission of digital data, bits per second for a serial line.

Calibration The act of certifying that a device, over its operating range, has a specified set of errors found from comparing it to a designated standard device.

Cavitation During the change from the vapor state to the liquid state of a fluid, cavitation may occur as pressure is increased. For example, as liquid flows through a restriction in a control valve, its velocity increases and its pressure decreases, which may cause some of the liquid to vaporize or flash into bubbles flowing in the remaining liquid. If the pressure recovery of the fluid downstream of the valve restriction raises the fluid pressure back above the vapor pressure, then the bubbles will collapse or implode, with possible cavitation damage to the valve body or piping.

Control mode *Manual mode* allows the human operator to manipulate the signal to the loop's process adjusting device. *Automatic mode* allows the operator to manipulate only the set point of the loop's controller. *Cascade mode* places the loop's set point under automatic adjustment by means of the output signal from another controller.

Control valve A process adjusting device that adjusts the flow of a fluid.

Controller Hardware or software logic that compares a measured process variable signal to a set-point signal and uses the result to calculate the controller output signal that adjusts the process adjusting device (see Figure 1.1).

Controller tuning constants The values of proportional action, integral action, and derivative action that are set into a controller to achieve a preferred response of its loop to set point or load disturbances.

Dead time The initial time during which the output signal of a process makes no response to its input signal. For example, the dead time of a conveyor belt is the time taken to detect a change in material flow rate at the exit of the belt after a change has been made to the material flow rate onto the entry to the belt.

Distributed control system A process control computer system that includes sensors, display consoles, and process adjusting devices distributed geographically throughout a factory.

Engineering units The units of measurement of the measured variable (for example, psia).

First-order lag time Same as time constant.

Flashing The change from liquid state to vapor state of a fluid as its pressure is reduced. For example, as liquid flows through a restriction in a control valve, its velocity increases and its pressure decreases, which may cause some of it to vaporize into bubbles flowing in the remaining liquid.

Frequency response The variation in gain [e.g., output sine wave amplitude (ma) / input sine wave amplitude (°C)] and the variation in phase shift (trigonometric degrees that the output sine wave lags or leads the input sine wave) of a device, over some band of frequencies of sine waves. For example, in Figure 2.19 the gain of a pneumatic-to-current converter decreases from 1.3333 to 0.0000130 ma/psi, and the phase angle decreases from 0 to –180° over a frequency band of 0.01–1000 radians/second.

Instrumentation symbols From ISA-S5.1 (see Appendix A).

Loop A group of devices organized for automatic control of a measured process variable (see Figure 1.1).

PLC Programmable logic controller replaces relay logic with a microprocessor and software.

Precision The minimum calibration error to which a device may be readily adjusted. For example, ±0.25°C in the linear relation of the output signal to the measured process variable.

Presentation of data The translation of transmitted sensor signals from the appropriate process variables into a format that conveniently displays for humans or (presents to machines) the changes of the process variables with time so that the status and trend of the process may be monitored continuously.

Process adjusting device A device, such as a control valve, that manipulates the process of a control loop so that the measured process variable approaches more closely the value of the controller set point.

Process adjustment Adjustment or manipulation of a process variable, usually a flow variable such as steam flow, to achieve a more desirable value for a measured process variable such as the temperature of a liquid in a tank.

Process control The action taken by a human or a machine to guide or to force the process to maintain a desirable state or change safely from an initial state to a more desirable state.

Process measurement Use of an instrument to sense the value of a process variable, such as temperature measured by a thermocouple, that usually transmits a signal proportional to this value to a remote location for process control.

Process variable A measurable physical or chemical characteristic of a process with its engineering units, such as temperature in °C.

psi Pounds per square inch of pressure.

psia Pounds per square inch absolute of pressure. Measures pressure difference relative to absolute zero pressure.

psid Pounds per square inch differential of pressure. Measures pressure difference between two points in a medium.

psig Pounds per square inch gauge of pressure. Measures pressure difference relative to atmospheric pressure.

psis Pounds per square inch standard of pressure. Measures pressure difference relative to 14.696 psia.

Range The minimum (zero) to maximum (zero + span) values for which an instrument is calibrated.

Ratholing May occur in a silo partially filled with solid particles, such that as particles are drawn out at the bottom, a cylindrical hole of the particles may be drawn out near the center so that the level at the top of the pile does not truly represent the quantity of particles in the silo.

Resolution The minimum detectable variation in the output signal caused by a corresponding change in the process variable (e.g., 0.05°C).

Scale The range seen on an indicating instrument.

Sensor A device that senses or detects the value of a process variable and generates a signal related to the value, such as an RTD (resistance temperature detector) that produces an electrical resistance that is related to its temperature and, as a consequence, senses the temperature of its surrounding medium. Additional transmitting hardware is required to convert the basic sensor signal to a standard transmission signal such as 4–20 ma. In this book the sensor is defined as the complete sensing and transmitting device.

Set point A value, entered by an operator into a controller, for which the controller adjusts its process to achieve a similar value of its measured process variable.

Span The difference between the maximum and minimum values to which an instrument is calibrated. For example, a pressure transmitter calibrated for 3–15 psig has a span of 12 psi (15 – 3 = 12).

Span adjustability The ability to adjust the range of values of the process variable that correspond to the output signal range. For example, the output range of 4–20 ma has a span of 16 ma that may be set to correspond to a minimum input span of 5°C or up to a maximum input span of 250°C.

Stability Maximum expected drift in calibration error with time, such as 0.1°C per six months.

Time constant The time for the output signal of a first-order process to respond to 63.2% of a step change in the input signal. For example, a voltage, V, is applied across a series combination of a capacitor, C, and a resistor, R. If the voltage on the capacitor is steady at V, and a voltage step change, ΔV, is made to V, then the time for the voltage, V_c, across the capacitor to increase by 63.2% of ΔV is the time constant. The input signal is V, and the output signal is V_c.

Transducer A device that converts a signal in one medium to a related signal in another medium. For example, an I/P transducer converts 4–20 ma to 3–15 psig.

Transmitter A device that is associated with a sensor and transmits a signal in a standard format related to the sensed process variable. For example, a pressure transmitter sends a signal in either 4–20 ma or 3–15 psig. A thermocouple is not a temperature transmitter, but it can be the basic sensor part associated with a transmitter that produces 4–20 ma.

Turndown ratio Ratio of maximum adjustable value to minimum adjustable value, such as a span turndown ratio of 10:1.

UPS (uninterruptible power supply) A group of batteries, charged continuously by the local AC main power, that supplies a high priority load (usually an AC load via an inverter). If the local AC main power fails, then the load continues to receive power automatically from the batteries for a certain period of time, usually long enough to shut down the process safely.

Zero The minimum value for which an instrument is calibrated.

Zero adjustability Suppression or elevation of the output zero signal relative to the process variable zero. For example, the process variable value of zero can be suppressed below the output zero of 4 ma or elevated so that it corresponds to an output greater than the output zero of 4 ma, which represents a negative value.

INDEX